MATHEMATICAL TEASERS

the text of this book is printed on 100% recycled paper

ABOUT THE AUTHOR

Professor Mira holds the degree of bachelor of science from Pennsylvania Military College and the degree of master of arts from Columbia University. After a few years in industry as an engineer, he began to teach mathematics at Manhattanville College, where he has been teaching for nearly 30 years. Professor Mira is now chairman of the mathematics department at Fort Lauderdale University. He has also been a lecturer in mathematics at Saint John's University and at the City University of New York. Mr. Mira is a fellow of the American Association for the Advancement of Science, and is author of *Arithmetic Clear and Simple* and coauthor of *Business Mathematics* and *Mathematics of Finance.*

EVERYDAY HANDBOOKS

MATHEMATICAL TEASERS

Julio A. Mira

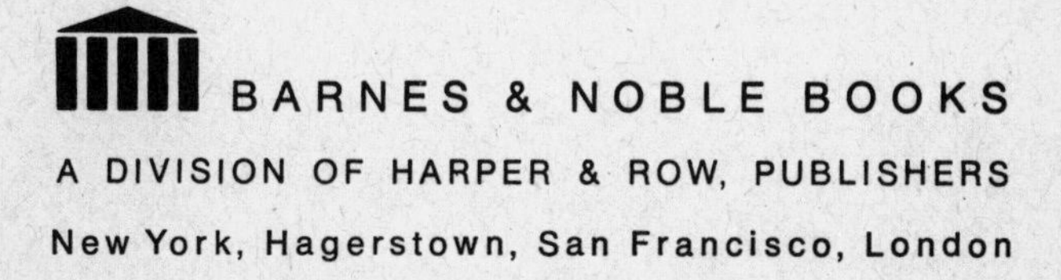
BARNES & NOBLE BOOKS
A DIVISION OF HARPER & ROW, PUBLISHERS
New York, Hagerstown, San Francisco, London

By BARNES & NOBLE, Inc.

L. C. Catalogue Card Number: 74–101122

SBN 389 00261 5

Printed in the United States of America

79 80 12 11 10 9

PREFACE

Solving mathematical puzzles has been a popular pastime for men since antiquity. From the cryptic utterances of the oracle of Apollo at Delphi in ancient Greece, six centuries before Christ, up to the present time, the solution of puzzles has contributed much to the development of modern mathematics. Three famous problems of antiquity: the squaring of the circle, the duplication of the cube, and the trisection of certain angles, have occupied Greek geometricians for centuries and have led to the discovery of many theorems and mathematical processes. In fact, only in recent times, with the development of the theory of groups, has it been proved that these problems cannot be solved by the formal methods of euclidean geometry. The problem of settling stakes in a game of chance, which was proposed to Pascal (1623–1662) by de Méré, led Pascal and Fermat (1601–1655) to formulate the fundamental principles of probability, a branch of mathematics, which is the foundation of modern statistical methods, and upon which the insurance business is based. Laplace (1749–1827) once said of probability, "A science which began with the consideration of play has risen to the most important objects of human knowledge."

Great thinkers like Anaxagoras (c.500–c.428 B.C.), Kepler (1571–1630), Leibnitz (1646–1716), Euler (1707–1783), Lagrange (1736–1813), and many others have devoted much of their time to solving mathematical puzzles. It was Leibnitz who remarked, "Men are never so ingenious as when they are inventing games."

This book contains a number of interesting, and often challenging, mathematical puzzles, which have been very carefully selected and put together especially for enjoyment by the person who may be neither a mathematical genius nor even a mathematician. It is the express hope of the author that the reader will enjoy these "mathematical teasers" as much as he has. The solution to each of the problems is found at the end of the chapter containing that particular problem. Some of the theory used in solving these problems can be found in *Arithmetic Clear and Simple,* also by the author.

The author wishes to express his gratitude to the heirs of Samuel I. Jones for permission to use his outstanding and unusual collection of mathematical puzzles, to Jonatha Foster for typing the manuscript, and to Pamela Singleton for her excellent cartoons.

TABLE OF CONTENTS

A = πr²
r

chapter 1
teasers for all

1. how far is near

A train leaves New York for Chicago traveling at the rate of 100 miles an hour. Another train leaves Chicago for New York an hour later traveling at the rate of 75 miles an hour. When the two trains meet, which one is nearer to New York?

2. the bank teller's problem

A man having $50 in a bank withdrew it as follows:

$20	leaving	$30
15	leaving	15
9	leaving	6
6	leaving	0
——		——
$50		$51

Where did the extra dollar come from?

3. yes, we have no apples

A girl with blue eyes
Went out to view the skies.
She saw an apple tree with apples on it.
She neither took apples nor left apples.
How many apples were on the tree?

4. the new arithmetic

Show that half of twelve is seven.

5. the calculating cook

If it takes 3 minutes to boil an egg, how long does it take to boil six eggs?

6. when larger is smaller

Which numbers, when divided by themselves, become larger than when multiplied by themselves?

7. square geometry

In this arrangement there are ten matches forming three squares. Which two must be removed to obtain two squares?

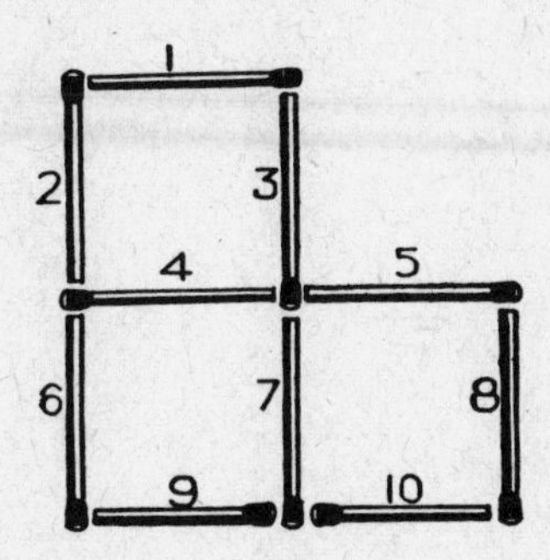

8. the careless clerk

A clerk sold 20 feet of wire and measured it out with a yardstick that was 3 inches too short. How much wire did the customer lack?

9. the tramp's cigarette

A tramp knows from experience that there is enough tobacco in four cigarette butts to make one cigarette.

How many cigarettes could he make from 29 cigarette butts?

10. discomposing

Which of the ten discs, at the left, must be moved to form the arrangement at the right by replacing only 3 discs?

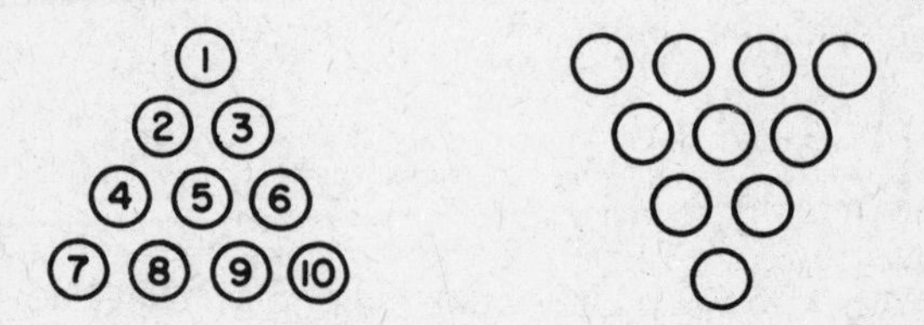

11. a melon drama

A farmer surrounded his square melon patch with a ditch 10 feet wide by 20 feet deep. He filled this ditch with water and considered the patch well protected against thieves. However, with only two $9\frac{1}{2}$-foot planks, thieves bridged the ditch and stole the melons. How were they able to do this?

12. even from odd

Write an even number using only odd digits, 1, 3, 5, 7, . . . , etc.

13. the land of ifthen

If five times four is thirty three,
What will the fourth of twenty be?

14. a bottle full equals a bottle empty

If a bottle half full equals a bottle half empty, and twice a half equals a whole, then a full bottle equals an empty bottle. Explain!

15. a dirt full

How many cubic inches of dirt are there in a hole 1 foot wide by 1 foot long by 1 foot deep?

16. tent station

An Arab sheik ordered his men to pitch his square tent within four trees which formed the corners of a square. When the sheik saw that the tent was too small, he told the men to pitch another square tent with twice the floor space of the original tent within the same four trees. How was this done?

17. how far does the horsefly fly?

Two horsemen, $\frac{1}{2}$ mile apart, are approaching each other. One of the horsemen travels at the rate of 6 miles an hour, while the other travels at the rate of 4 miles an hour. A horsefly traveling at the rate of 20 miles an hour flies back and forth between the two horses until they meet. How far does the horsefly fly?

18. milady's bracelet

Find the minimum cost of making six pieces of chain containing five links each into a circular chain containing thirty links, if it costs 2¢ to open a link and 3¢ to weld a link.

19. nail juggling

Transform this fraction into a fraction whose value equals one third by replacing only one nail.

20. identity not lost

Write the number 20 using four nines.

21. the lazy monkey

Two monkeys are balanced on a rope which goes over a pulley. The lazy monkey remains stationary, while the other one climbs up the rope. What happens to the lazy monkey?

22. the rook's swindle

A man purchased a pair of shoes that cost $25 and gave the merchant a $100 bill. After the man had gone with his shoes and his change, the merchant took the $100 bill to the bank where he was told that the bill was a counterfeit. What was the total loss to the merchant?

23. johnson's cat

Johnson's cat went up a tree,
Which was sixty feet and three.
Every day she climbed eleven;
Every night she came down seven.
Tell me, if she did not drop,
When her paws would touch the top!

24. another identity not lost

Express the number 3 using three threes.

25. the matchmaker

In this arrangement there are sixteen matches. Remove four and replace one so that you can spell the thing out of which good matches are made!

26. hotel stretch

Eight men went to a hotel and asked for separate rooms. The manager told them that only seven rooms were available, but that he would give each of them a separate room anyway. He put two men into the first room, the third man into the second room, the fourth man into the third room, the fifth man into the fourth room, the sixth man into the fifth room, and the seventh man into the sixth room. He then went back to the first room and put one of the two men there into room number seven. Thus, he put eight men into seven rooms, giving each one of them a separate room. What's wrong with this story?

27. the road to st. ives

As I was going to St. Ives,
I met a man with seven wives.
Every wife had seven sacks,
Every sack had seven cats,
Every cat had seven kits.
Kits, cats, sacks, and wives,
How many were going to St. Ives?

28. a good match

Here is an incorrect equation. Change it into a correct one by replacing just one match.

29. take it away and it's there

Subtract 45 from 45 and obtain 45 as the difference.

30. square eggs

If feet multiplied by feet equals square feet, then eggs multiplied by eggs equals square eggs. What's wrong with this statement?

31. some more squares

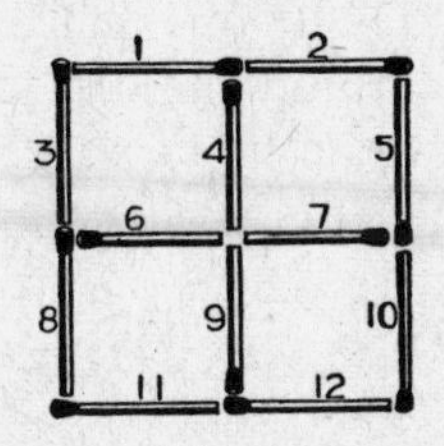

In this arrangement there are twelve matches. Which two must be removed to get two squares?

32. things are not what they seem

Even when figures are drawn very accurately, it is possible to be deceived by appearances. Which line is longer, *AB* or *CD*? Will lines *EF* and *GH* ever meet? Is line *JK* higher or lower than line *NO*?

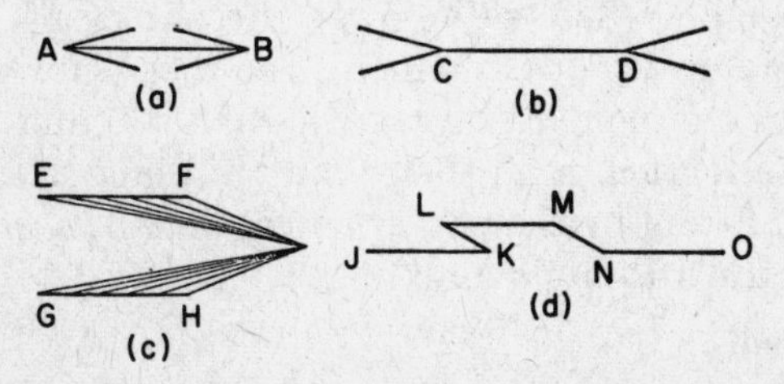

33. travel in flatland

Trace this diagram with a pencil, connecting all the nine dots with four straight lines, but do not lift your pencil.

34. of cats, dogs, and mice

A man has a dog, a cat, and a mouse. He must transport them across a river one by one, since the dog cannot be left alone with the cat and the cat cannot be left alone with the mouse. How can he do this?

35. when twice is thrice

Thrice what number is twice that number?

36. the light-fingered porter

Three men traveling by automobile stopped at a motel to get a room for the night. The manager told them that only one \$30 room with three beds was available. The men agreed to share the cost of the room equally. In the meantime, the manager realized that the room rented for only \$25, so he gave the porter \$5 and told him to divide it equally among the men. However, the light-fingered porter gave each man \$1 and kept \$2 for himself. Each man then paid \$9. Since $3 \times 9 = 27$ plus the \$2 kept by the porter leaves only \$29, what happened to the missing dollar?

37. simple addition

Show that eleven plus two makes one.

38. for the postmaster

If thirteen stamps cost a cent and a quarter, how much does a stamp cost?

39. for the english scholar

Which is correct: 9 and 6 are 14, or 9 and 6 is 14?

40. old bills for new ones

Which would you prefer, an old $10 bill or a new one?

41. the mathematical bookworm

An ambitious "bookworm" decided to digest some mathematics from a two-volume treatise on probability placed on a shelf in the usual manner. Beginning with the first page of volume I, the bookworm bored in a straight line through to the last page of volume II at the rate of $\frac{1}{2}$ inch per day. If the pages of each volume are 1 inch thick and each cover is $\frac{1}{8}$ inch thick, how long did it take the bookworm to digest all that mathematical knowledge?

42. two to two

Write the number 2 using seven twos.

43. a party with threes

Find the value of $3 + 3 \times 3 - 3 \div 3 - 3$.

44. how fast is the cutter?

How long will it take to cut a 60-yard piece of cloth into 1-yard pieces, allowing only 1 minute for each cut?

45. five squares from three

In this arrangement there are twelve matches forming three squares. Which three must be replaced to get five squares?

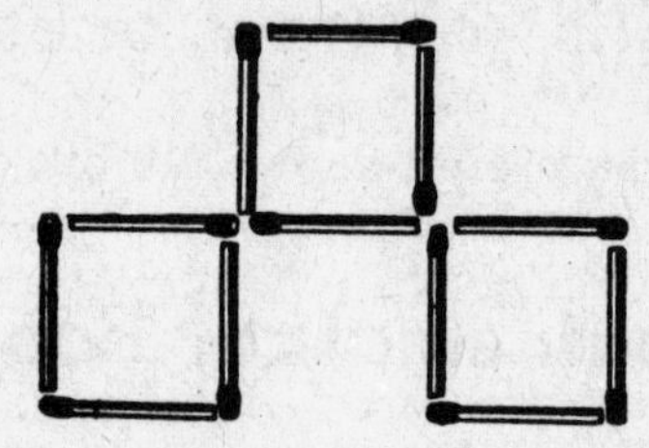

46. a fast buck

The hunt is on, the trumpets blare,
The hounds are chasing the buck for fair.
Then into the woods the buck does run
At thirty miles an hour, not for fun.
The woods are dense and sixty miles wide.
Now I'd have you take a stand upon,
How long the buck into the woods did run?

47. a perplexing choice

A young man received two offers: Company A offered him $8,000 a year with a $300 raise every 6 months;

Company B offered him \$8,000 a year with a \$1,200 raise every 12 months. Which offer should he accept?

48. the eager beaver

The young man in the preceding puzzle, having made the right choice, became overambitious and immediately asked for a raise. His boss refused to give it to him and showed him that he was being paid for not working by making the following claim:

There are 365 days in the year.
You sleep 8 hours a day, which is one third of the time, or 122 days; this leaves $365 - 122 = 243$ days.
You spend 8 hours each day, or 122 days, in rest and recreation; this leaves $243 - 122 = 121$ days.
You do not work on Saturdays and Sundays, or 104 days; this leaves $121 - 104 = 17$ days.
You have 2 weeks vacation, or 14 days; this leaves $17 - 14 = 3$ days.
You get the Fourth of July, New Year's Day, and Christmas Day; this leaves $3 - 3 = 0$.
In fact, you owe the company time, since you also get 1 hour off for lunch, Washington's Birthday, Thanksgiving Day, and Labor Day.
Explain the fallacy of this claim.

49. grandpa's watch chain

Find the minimum cost of making five pieces of chain with three links each into a straight watch chain for grandpa, if it costs 2¢ to open a link and 8¢ to weld a link.

50. triality

Using six matches of equal length, form four equilateral triangles.

51. the pigpens

A father offers his son 21 pigs for his very own, provided the son build four pigpens so that each pen encloses an even number of pairs of pigs and an odd pig besides. How can the son do this?

52. post time

A man has a square plot of ground with twelve posts equally spaced on each side. How many posts are required altogether?

53. rhythmic roman

A Roman boy was given this puzzle:

> Take one hundred one, and to it affix
> The half of a dozen, or if you please six;
> Put fifty to this, and then you will see
> What ev'ry good boy to others should be.

54. daft arithmetic

Show that half of nine is four.

55. square quarters

How many quarter-inch squares does it take to make an inch square? How many quarter-inch cubes does it take to make an inch cube?

56. dozens of dozens

Which, if either, is the greater, six dozen dozen or a half a dozen dozen?

57. show biz

A showman is traveling with a wolf, a goat, and a basket of cabbages. He must transport them across a river one by one, since the wolf cannot be left alone with the goat and the goat cannot be left alone with the cabbages. The only means of transportation is a small boat that will accommodate only him and one other at one time. How can he transport them all across the river?

58. match wit

Enclose three equal squares with eight matches, four of which are exactly half the length of the others.

59. post man

A man having a garden 10 rods square, wishes to know how many posts will be required to enclose his land, if the posts are placed exactly 1 rod apart!

60. more triality

How many equilateral triangles can be formed with nine matches of equal length?

61. juicy fruits

There are six oranges in a box. Without cutting any of the oranges, divide them among six boys in such a way that one orange is left in the box.

62. duck a duck

A duck before two ducks, a duck behind two ducks, and a duck in the middle. How many ducks are there?

63. lend an ear

There are 21 ears of corn in a hollow stump. How long will it take a squirrel to remove all of them, if he carries off three ears every day?

64. yard tract

What part of $\frac{1}{2}$ square yard is $\frac{1}{2}$ yard square?

65. common places

Write the number 100 using arabic numerals, but do not use any zeros.

66. more daft arithmetic

Show that six and six equals eleven.

67. slippery snail

If a snail, crawling up a pole 10 feet high, climbs 3 feet each day and slips back 2 feet each night, how long will it take the snail to reach the top?

68. rolling around

If two coins the same size are placed on a table so that they are touching at one point, and one of the coins is rolled around the other, how many revolutions must the rolled coin make in order to return to its original position?

69. the millionaire

A man owns a triangular lot in Times Square, New York City. Its dimensions are exactly 101 feet by $51\frac{1}{2}$ feet by $49\frac{1}{2}$ feet. He guarantees to have just what the deed calls for, and he will reward the first person who finds the exact area of the lot with a 1% interest in the land. What is the area?

70. a count up

How many triangles does this figure contain? How many squares does it contain?

solutions

1. how far is near

Both trains are at the same distance from New York.

2. the bank teller's problem

If the final remainder were zero, the sum of the amounts subtracted would equal the original amount. However, the sum of the remainders does not equal the sum of the amounts subtracted, therefore the sum of the remainders does not yield the original amount.

3. yes, we have no apples

Two apples. She took one apple and left one apple.

4. the new arithmetic

Write 12=XII, remove the lower half ΛII, and you have VII=7.

5. the calculating cook

3 minutes, if all the eggs are boiled at the same time.

6. when larger is smaller

Any proper fraction. For example: $\frac{1}{3} \div \frac{1}{3} = 1$, but $\frac{1}{3} \times \frac{1}{3} = \frac{1}{9}$.

7. square geometry

Remove matches numbered 6 and 9.

8. the careless clerk

18 inches. Only 3 inches were lost on each yard in measuring 18 feet, or 6 yards, with the 33 inch stick, but nothing was lost in measuring the last two feet. Therefore, the total loss to the customer was $6 \times 3 = 18$ inches.

9. the tramp's cigarette

Nine. The 29 butts yield seven cigarettes with one butt left over. When the tramp smokes these seven cigarettes, he has seven new butts, and, with the butt he had left over, he now has eight butts from which he can make two cigarettes. The total number of cigarettes made by the tramp is seven plus two, or nine cigarettes, with two butts left over.

10. discomposing

Discs numbered 1, 7, and 10.

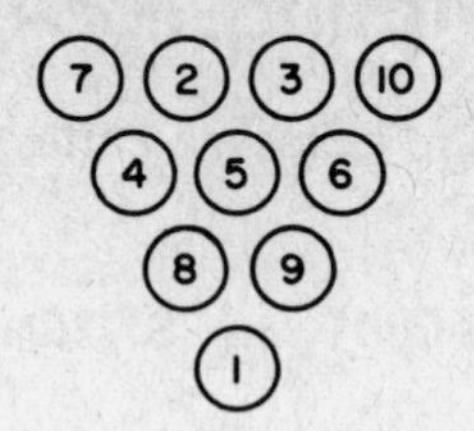

11. a melon drama

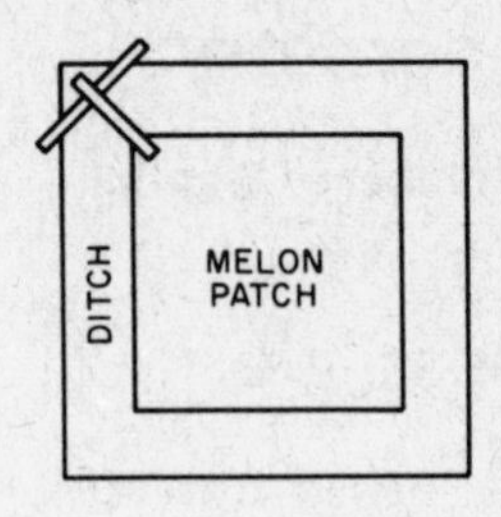

12. even from odd

$1\frac{1}{1} = 1 + 1 = 2$, $3\frac{3}{3} = 3 + 1 = 4$, $5\frac{5}{5} = 5 + 1 = 6$, etc.

13. the land of ifthen

$8\frac{1}{4}$, since $5 \times 4 = 20$, and $\frac{1}{4}$ of 20 is $\frac{1}{4}$ of 33, or $8\frac{1}{4}$.

14. a bottle full equals a bottle empty

If the problem refers to the capacity of the bottle, the conclusion is correct; but, if the problem refers to the amount of liquid in the bottle, then twice the amount of liquid in a bottle half empty is equal to twice the amount of liquid in a bottle half full. Therefore, you would have a full bottle and not an empty bottle.

15. a dirt full

None. There is no dirt in a hole.

16. tent station

The larger tent was pitched so that its base formed a square with E, F, G, H as the corners and A, B, C, and D as midpoints on the sides.

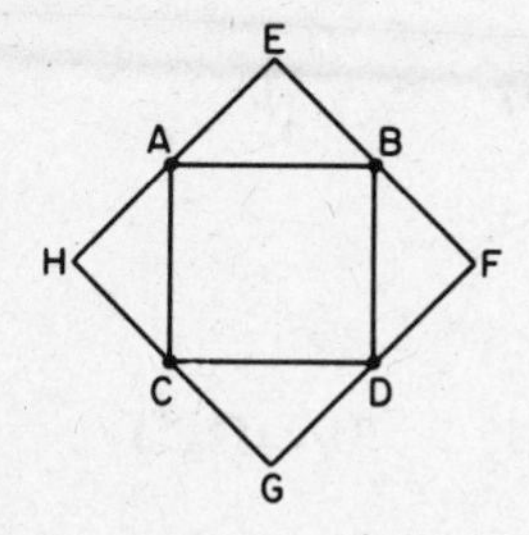

17. how far does the horsefly fly?

1 mile. Horsemen approach each other at $6 + 4 = 10$ miles per hour. The original distance between them is $\frac{1}{2}$ mile. Hence, they meet in $\frac{1}{2} \div 10 = \frac{1}{20}$ of an hour. Therefore, the horsefly flies for $\frac{1}{20}$ of an hour at 20 miles per hour, or $\frac{1}{20} \times 20 = 1$ mile.

18. milady's bracelet

25¢. Cut each of the five links of one of the pieces of chain. Use four links to connect the five remaining pieces of chain and the fifth link to connect the ends. Since there are a total of 5 cuts and 5 welds, it costs $5 \times 2¢ + 5 \times 3¢ = 25¢$.

19. nail juggling

$$\frac{II}{VI} = \frac{2}{6} = \frac{1}{3}$$

20. identity not lost

$9 + \frac{99}{9}$.

21. the lazy monkey

The lazy monkey rises.

22. the rook's swindle

$75 and the shoes.

23. johnson's cat

The 14th day. The cat's daily upward gain is $11 - 7 = 4$ feet. In 13 days she climbs $13 \times 4 = 52$ feet up the tree, and on the 14th day, she climbs 11 feet. Since $52 + 11 = 63$ feet, the cat reaches the top on the 14th day.

24. another identity not lost

$3 \times \frac{3}{3}$ or $\sqrt[3]{3^3}$.

25. the matchmaker

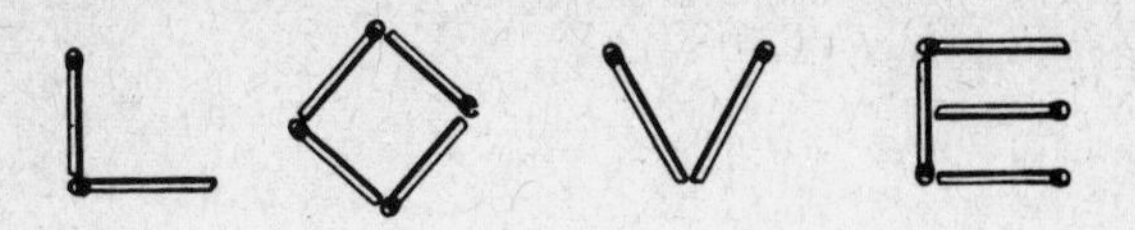

26. hotel stretch

The manager did not give the eighth man a room.

27. the road to st. ives

One. I am the only one going to St. Ives.

28. a good match

29. take it away and it's there

$$\begin{array}{r} 9+8+7+6+5+4+3+2+1=45 \\ 1+2+3+4+5+6+7+8+9=45 \\ \hline 8+6+4+1+9+7+5+3+2=45 \end{array}$$

30. square eggs

A multiplier is always an abstract number. In finding areas, for example, the statement, "feet times feet equals square feet," is misleading. The fact is that in finding the area of a rectangle, we are not multiplying feet by feet, which is absurd, but the number of square feet in the base by the number of rows, which is an abstract number.

31. some more squares

Remove either of the following pairs of matches: numbers 4 and 6, 4 and 7, 6 and 9, or 7 and 9.

32. things are not what they seem

AB = CD; EF is parallel to GH, therefore these lines will never meet; lines JK and NO are at the same level.

33. travel in flatland

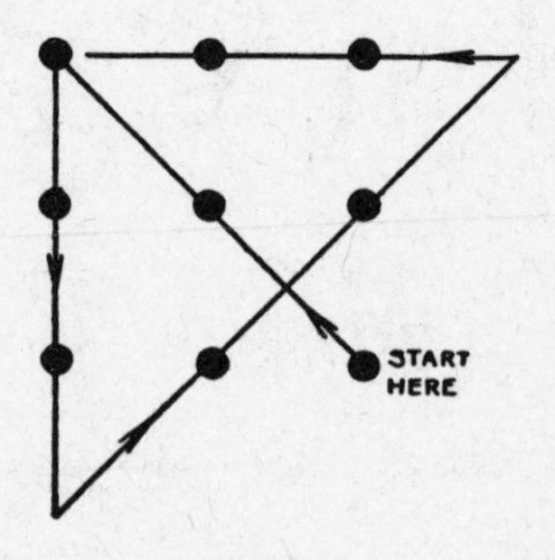

34. of cats, dogs, and mice

First, he takes the cat across. He returns and takes the mouse across and brings back the cat; then he takes the dog across and returns and takes the cat across.

35. when twice is thrice

Zero.

36. the light-fingered porter

Each man paid $10 and should have gotten back a third of $25. Thus, $\$10 - \frac{\$1}{3} \times 5 = \$10 - \frac{\$5}{3} = \frac{\$25}{3} = \$8\frac{1}{3}$ is the actual cost of the room per man. Therefore, the calculation should be $8\frac{1}{3} \times \$3 = \$25$. Then, $25 + $3, which the porter returned, plus $2, the porter's profit, is $\$25 + \$5 = \$30$.

37. simple addition

11 o'clock plus 2 hours is 1 o'clock.

38. for the postmaster

2¢. The cost of thirteen stamps is 1¢ and a quarter, or 26¢; hence one stamp costs 2¢.

39. for the english scholar

Neither. Since $9 + 6 = 15$; but both statements are grammatically correct.

40. old bills for new ones

An old $10 bill is usually preferable to a new $1 bill.

41. the mathematical bookworm

$\frac{1}{2}$ day. Since usually volume I is to the left of volume II, the first page of volume I is separated from the last page of volume II by only two covers, which together total $\frac{1}{4}$ inch. Since the bookworm bored through at the rate of $\frac{1}{2}$ inch per day, it took him $\frac{1}{4} \div \frac{1}{2} = \frac{1}{2}$ day.

42. two to two

$2 + \frac{22}{22} - \frac{2}{2}$.

43. a party with threes

In a series of indicated operations, multiplications and divisions are to be performed first in the order in which they appear from left to right; additions and subtractions are to be performed after these operations have been completed. Hence, $3 + 3 \times 3 - 3 \div 3 - 3 = 3 + 9 - 1 - 3 = 8$.

44. how fast is the cutter

59 minutes. The last piece does not have to be cut.

45. five squares from three

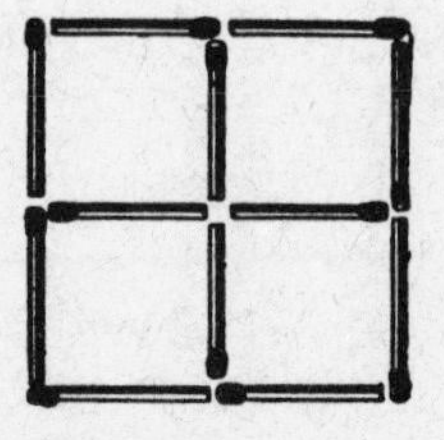

46. a fast buck

1 hour. The width of the woods is 60 miles; hence the buck ran into the woods 30 miles, for after that he would be running out of the woods. Since the buck travels 30 miles per hour, he runs for 30 miles $\div$ 30 miles per hour $=$ 1 hour.

47. a perplexing choice

Company A. The young man would receive the following yearly amounts:

Year	Company A	Company B
1st	$4,000 + 4,300 = $8,300	$8,000
2nd	4,600 + 4,900 = 9,500	9,200
3rd	5,200 + 5,500 = 10,700	10,400
4th	5,800 + 6,100 = 11,900	11,600
.	.	.
.	.	.
.	.	.

Company A's offer is better by $300 a year.

48. the eager beaver

The company has a claim to the young man's working hours only, that is, to 8 hours a day for a 5-day week, discounting all legal holidays.

49. grandpa's watch chain

30¢. Cut each of three links of one of the pieces and connect the four other pieces. Since there is a total of three cuts and three welds, it costs $(3 \times 2¢) + (3 \times 8¢) = 30¢$.

50. triality

Form an equilateral triangle with three matches. With this triangle as the base, erect a pyramid with the three remaining matches. The three sides of the pyramid and the base should form four equilateral triangles.

51. the pigpens

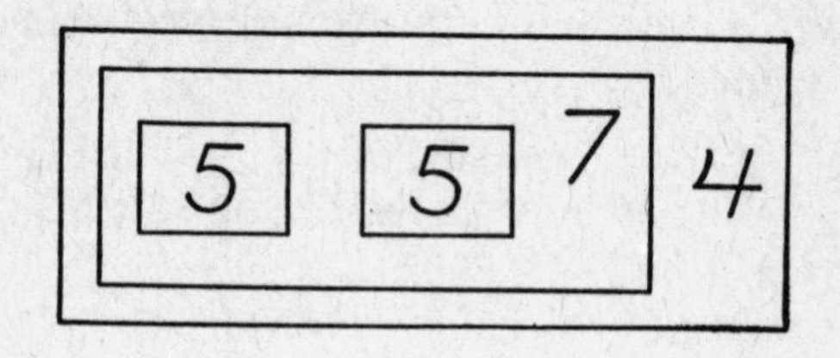

52. post time

44 posts.

53. rhythmic roman

One hundred one = CI
Six = VI
Fifty = L

Thus, we obtain CIVIL.

54. daft arithmetic

Write 9=IX , remove the lower half IΛ , and you have IV=4

55. square quarters

16 quarter-inch squares; 64 quarter-inch cubes.

56. dozens of dozens

Six dozen dozen $= 6 \times 12 \times 12 = 864$; one half a dozen dozen $= \frac{1}{2} \times 12 \times 12 = 72$.

57. show biz

He takes the goat across. He returns and takes the basket of cabbages across and brings back the goat; then he takes the wolf across and returns and takes the goat across.

58. match wit

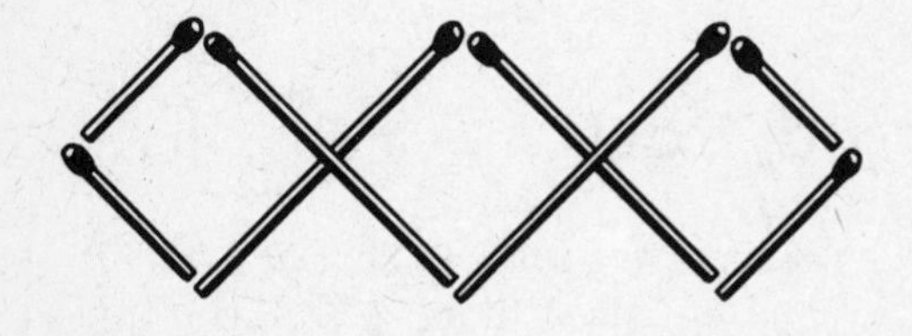

59. post man

40 posts.

60. more triality

Note that the perimeter of the figure also forms an equilateral triangle.

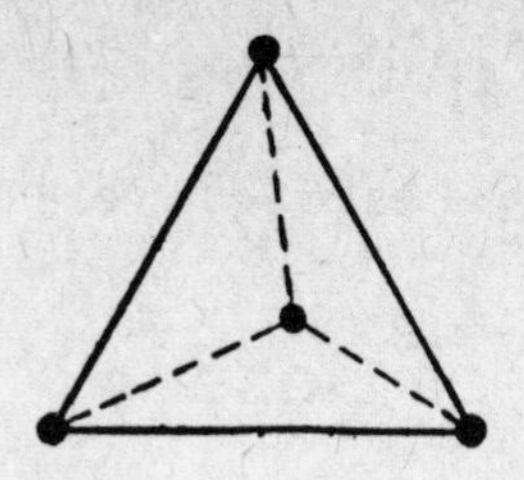

61. juicy fruits

Give one of the boys the box with an orange in it.

62. duck a duck

Three ducks.

63. lend an ear

21 days. Two of the ears are the squirrel's.

64. yard tract

Since $\frac{1}{2}$ yard square equals $\frac{1}{2} \times \frac{1}{2} = \frac{1}{4}$ square yard and $\frac{1}{4}$ square yard equals half of $\frac{1}{2}$ square yard, $\frac{1}{2}$ square yard equals half of $\frac{1}{2}$ yard square.

65. common places

$99\frac{2}{2} = 99\frac{3}{3} = 99\frac{4}{4} \ldots = 99 + 1 = 100.$

66. more daft arithmetic

Write 6=VI, turn VI upside down, ΛI, put it at the bottom of VI, and you have XI=II.

67. slippery snail

8 days. At the beginning of the eighth day the snail is 3 feet from the top, and it reaches the top at the end of that day.

68. rolling around

One revolution.

69. the millionaire

Zero area. These dimensions do not form a triangle but a line, since $51\frac{1}{2} + 49\frac{1}{2} = 101$. Therefore the lot has no area.

70. a count up

150 triangles; 31 squares, including the square encompassing the entire figure.

chapter 2
more teasers for all

1. discotheque go-go

The outer track of a long-playing record is $11\frac{1}{2}$ inches in diameter and the label at the center is $5\frac{1}{2}$ inches in diameter. If the record has twenty grooves per inch, how far does the needle travel to play the whole record?

2. instant product

Find the product of the digits 1, 2, 3, 4, 5, 6, 7, 8, 9, 0 as fast as possible, and time yourself.

3. an L of a design

A lady asked her son Henry, an architect, to design a house for her in the form of an "L" with windows in each wall and with all the windows having a southern exposure. How did the architect solve this problem?

4. a long count

How many triangles are there in this figure?

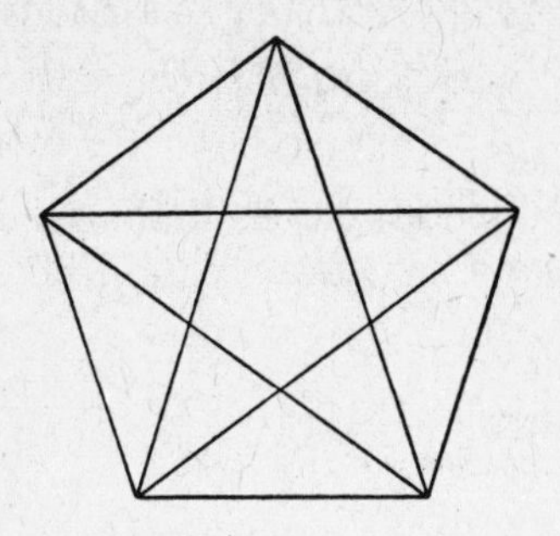

5. bells are ringing

If it takes a clock 6 seconds to strike 4 o'clock, how long will it take the clock to strike 12 o'clock?

6. after dinner

Using eight large toothpicks, form a regular polygon with more than four sides, two squares, and eight triangles.

7. paying off old scores

Three friends, John, Frank, and George, met one day. John owed Frank $5, Frank owed George $5, and George owed John $5. Each of them was feeling some embarrassment in the presence of his creditor, as Bill came over to them. John then asked Bill to lend him $5. Bill did so, and John immediately handed it to Frank, who, in turn, gave the $5 to George. George then paid John his $5. John then handed the $5 back to Bill with thanks. Thus, all debts were paid. Explain!

8. out on a limb

There were nine pigeons on the limb of a tree. A hunter shot three of them. How many remained?

9. take it away

From six you take nine
And from nine you take ten,
Then from forty take fifty
And six will remain.

10. traveling in flatland

Trace this diagram without lifting your pencil, folding the paper, or going over any line more than once.

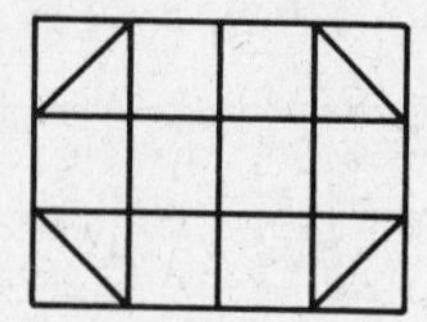

11. mad arithmetic

Show that half of thirteen is eight.

12. how come?

Two trains left Nashville for Louisville at the same time traveling on parallel tracks. Both arrived at Louisville at the same time. One train made the trip in 80 minutes, while it took the other train 1 hour, 20 minutes. Explain!

13. shuttle

The shuttle service has a train going from Washington to New York City and from New York City to Washington every hour on the hour. The trip from one city to the other takes $4\frac{1}{2}$ hours, and all trains travel at the same speed. How many trains will you pass in going from Washington to New York City?

14. touching matches

Arrange six matches so that each match touches four others.

15. get rich quick

A man with $1 wanted $1.25, so he pawned the $1 for 75¢ and then sold the pawn ticket for 50¢. He then had $1.25. Who lost in this transaction?

16. playing with matches

Arrange six matches so that every match touches every other match.

17. gluttonous cats

If six cats eat six rats in 6 minutes, how many cats will it take to eat a hundred rats in 100 minutes at the same rate?

18. a lot of bull

If a bull standing on three legs weighs 1000 pounds, how much will this bull weigh when standing on four legs?

19. an odd one

Which numbers when divided by 2 will have 1 left over?

20. triangulate

How many equilateral triangles can be formed from nine matches of equal length?

21. strike this one out

To one half of a strike
Add two thirds of a ton;
Then a stone you will see,
If 'tis properly done.

22. even off

Take one from seven, two from eleven, and come out even.

23. something different

What is the difference between twice twenty five and twice five and twenty?

24. playing post office

If it takes twelve 1¢ stamps to make a dozen, how many 2¢ stamps does it take?

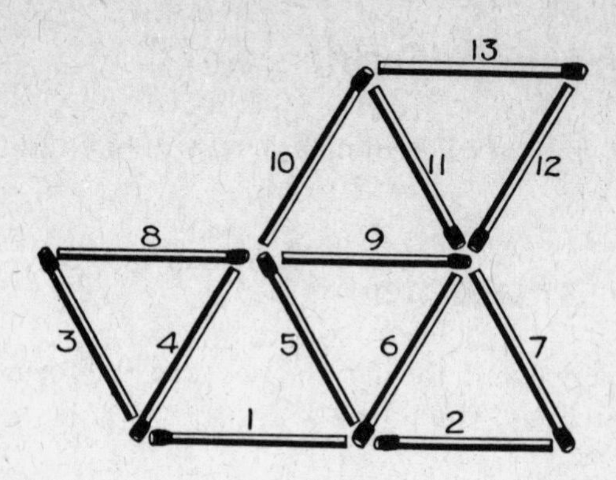

25. six yield three

In this arrangement there are thirteen matches forming six equilateral triangles. Which three matches must be removed to get three triangles?

26. bye bye blackbirds

Twice four and twenty blackbirds
Were sitting in the rain;
I shot and killed a seventh part,
How many did remain?

27. too touching

Can you place four coins on a table all at equal distances?

28. no lost marbles

Can you place four marbles on a table all at equal distances?

29. same but different

Write the number 30 with three digits all alike.

30. two for the summing

Add five digits, all alike, so that their sum is 14.

31. it all adds up

Add 1, 2, 3, 4, 5, 6, and 7, written in order so that they add up to 100.

32. figure this one out

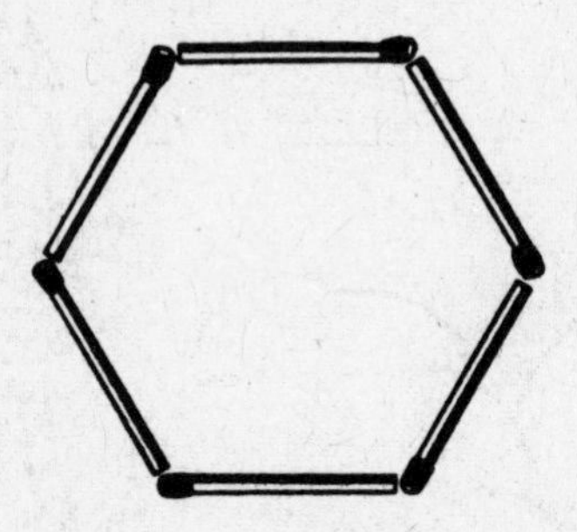

In this arrangement there are six matches forming a regular hexagon. Take three more matches and arrange all nine so that they form another six-sided figure.

33. double up

What coin doubles its value when its half is removed?

34. lover boy

You, sir, I ask to plant a grove
To show that I'm your lady love.

This grove though small must be composed
Of twenty-five trees in twelve straight rows.
In each row five trees you must place,
Or you shall never see my face.

35. lover boy does it again

You, sir, who did plant the grove
To show that I'm your lady love.
Must now plant within a square
A smaller grove of nine trees rare.
In each of ten rows three trees you must place,
And then you may forever see my face.

36. weigh carefully

Which is heavier, a pound of gold or a pound of feathers?

37. water boy

A boy went to a spring to get exactly 4 quarts of water, but he had only two jars, one jar holding 5 quarts and the other jar holding 3 quarts. How was the boy able to get the 4 quarts using only these jars?

38. take away and get more

Show that 1 taken away from nineteen leaves twenty.

39. equal but not the same

If a quart is $\frac{1}{4}$ gallon and also $\frac{1}{8}$ peck, why is a gallon not equal to $\frac{1}{2}$ peck?

40. crack this bottle

If a bottle and a cork together cost $1.10, and the bottle costs $1 more than the cork, how much does each cost?

41. mental exercise

Find the sum of the numbers from 1 to 50, inclusive, without adding them or using a formula.

42. letter perfect

Add one third of twelve to four fifths of seven so as to get eleven.

43. a re-match

Change this incorrect equation, to a correct one by replacing just one match.

44. nutty arithmetic

Show that four and four is nine.

45. a neat profit

How much is ten thousand percent of a cent?

46. a-hunting we will go

As I was traveling on the forest grounds,
Up starts a hare before my two greyhounds.
The dogs being light of foot they fairly run,
Unto his 16 rods just 21.
The distance he started up before,
Was four score sixteen rods just, and no more.
Now, I'd have you unto me declare
How far they ran before they caught the hare.

47. good neighbor policy

Three brothers were left an estate consisting of seventeen cows which were to be divided as follows: the oldest brother was to receive one half of the cows, the next oldest one third, and the youngest one ninth. The brothers wished to divide the cows among themselves without slaughtering or selling any of them. While they were discussing this subject, a neighbor offered to solve the problem by adding one of his cows to the seventeen. The oldest brother would then receive nine cows, the next oldest six cows and the youngest two cows. The neighbor asked if they were satisfied and they replied that they were, so he said that since the sum of nine, six, and two amounted to only seventeen cows, his cow was left over and he will take her back. Explain!

48. more nutty arithmetic

Show that one added to nine makes twenty.

49. a horse laugh at lots of corn

If a hundred horses eat 100 tons of corn in 100 days, how many tons of corn will ten horses eat in 10 days?

50. lad and dad

This illustration shows a boy and his father. Which one is taller?

51. apple sauce

If one third of six apples sells for 3¢, what will one fourth of forty apples sell for?

52. medium to rare

I bought a steak that weighed 5 pounds plus half of its weight. How much did this steak weigh?

53. duck this one

Thrice as many ducks are three more than twice as many. How many ducks are there?

54. get to the root of this one

Change this incorrect equation to a correct one by replacing just one match.

55. the jealous husbands

Three jealous husbands, traveling with their wives, found it necessary to cross a stream in a boat which held only two persons. Each of the husbands refused to let his wife cross with either of the other male members of the party, unless he himself was also present. How was the passage arranged?

56. the old and the new

Five times seven and seven times three
Add to my age and it will be
As far above six nines and four
As twice my years exceeds a score.

57. all lines are equal

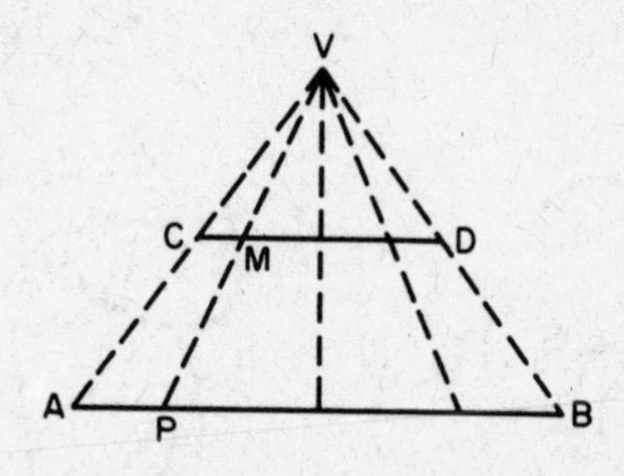

Given any two lines AB and CD, draw two lines passing through the end points A and C, and B and D meeting at the point V. From every point on AB, a line may be drawn to V. Every one of these lines will also cut CD at a point. For example, the line VP will cut CD at M. This is easy to visualize, as follows: Hold a thin piece of thread so that one end is at V with the other end at B. Then, keeping the end at V fixed, and holding the thread taut, move the free end at B toward A. Since every one of these lines cuts CD, and a line has no thickness, for every point on AB there will be a corresponding point on CD. Hence, CD has exactly the same number of points as AB, and is therefore equal to it in length. Explain!

58. the student's blunder

In solving a problem a student mistakenly divided instead of multiplying a number by 6, obtaining 15 as a result. Find the number and give the correct answer.

59. a fish story

Ten fish I caught without an eye,
And nine without a tail;
Six had no head, and half of eight
I weighed upon the scale.
Now who can tell me as I ask it,
How many fish were in my basket?

60. writer's cramp

Write the number 55, using five fours.

61. ancient write-in

Express the number 100 using six matches.

62. sweet in his cups

Divide ten pieces of sugar among three cups, without breaking the pieces, so that every cup contains an odd number of pieces.

63. stu simple

Senators Stu Simon and Lezan Halfbright were told to divide $28 million among seven nations for the dental care of weak toothed aardwolves. Stu Simon made the division as follows:

$$\begin{array}{r} 13 \\ 7\overline{)28} \\ \underline{7} \\ 21 \end{array}$$

He explained this by saying that 7 did not go into 2, so he put the 2 down below. Since 7 went into 8 once and 2 plus the 1 remaining made 3, each nation was entitled to $13 million.

The result did not seem quite right to Halfbright, so in order to prove it, he put the number 13 down in a column seven times. Then he added up all the 1's and all the 3's; this made 28, proving Simon's result as shown at the left. But then Simon said to Halfbright, "There is no need to check it this way. I can do it faster by multiplying 13 by 7 this way."

$$\begin{array}{c} 13 \\ 13 \\ 13 \\ 13 \\ 13 \\ 13 \\ 13 \\ \hline 7 + 21 = 28 \end{array}$$

$$\begin{array}{r} 13 \\ \underline{7} \\ 21 \\ \underline{7} \\ 28 \end{array}$$

"As you see," continued Simon, "seven times three equals twenty-one and seven times one equals seven, then twenty-one plus seven equals twenty-eight, which shows that we are both better mathematicians than Pit Agoras."

64. nothing to it

Show that half of 888 is twice as much as nothing.

65. too many eggs in one basket

Two farmers were selling eggs. Each had 300 eggs. The first farmer was asking 1¢ for two eggs, while the second farmer was asking 1¢ for three. They agreed to put their eggs together and sell five eggs for 2¢; but 10¢ was lost in the transaction. Explain!

66. matchless figures

In this arrangement there are thirteen matches. Which four matches must be removed to get two triangles, a trapezoid, and a parallelogram?

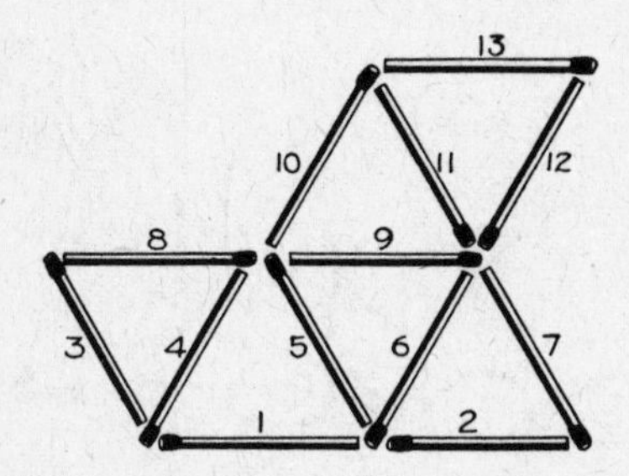

67. grouping

Add four two's and get five for the sum.

68. cut it out

Divide a pie into eight equal pieces with exactly three cuts.

69. use a fourth dimension

Put a dollar bill into a bottle, tightly cork the bottle, and remove the dollar bill without pulling out the cork or breaking the bottle.

70. the tender missionaries and the hungry cannibals

Three missionaries and three cannibals have to cross a river. They can't swim, but have a small boat with an outboard motor which will carry only two persons. Each of the missionaries can operate the outboard motor, but only one of the cannibals can do so. A missionary can be alone with a cannibal, but if one missionary is left with two cannibals, the missionary will be immediately killed and eaten by the cannibals. How can they cross the river so that all six are safely delivered on the other side of the river?

solutions

1. discotheque go-go

6 in. To play one side of the record, the needle travels $(11\frac{1}{2} - 5\frac{1}{2}) \div 2 = 3$ inches. To play the whole record, the needle travels $3 \times 2 = 6$ inches.

2. instant product

Zero. Any number multipled by zero equals zero.

3. an L of a design

The architect chose the north pole as the site for the house.

4. a long count

Thirty-five triangles.

5. bells are ringing

22 seconds. In making four strokes, there are three intervals; therefore, in a 6 second period, each interval lasts 2 seconds, and in making twelve strokes, there are eleven 2 second intervals.

6. after dinner

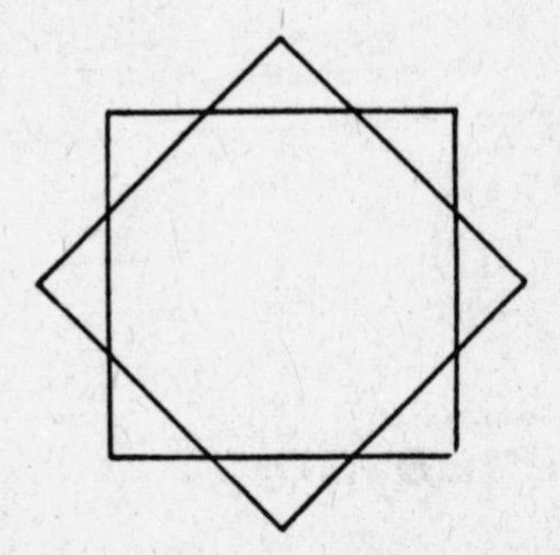

7. paying off old scores

Since the $5 made a full circle, all debts are cancelled; therefore Bill's contribution was not necessary.

8. out on a limb

The three that were shot. The rest flew away.

9. take it away

```
SIX    IX    XL
 IX     X     L
———   ———   ———
S      I     X
```

10. traveling in flatland

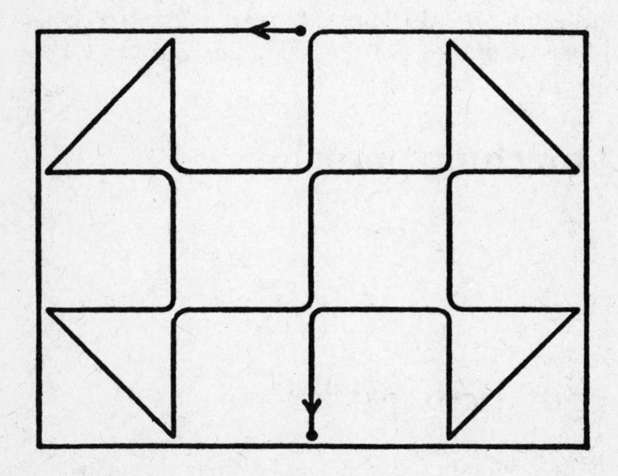

11. mad arithmetic

Write 13=XIII, remove the lower half ΛIII, and you have VIII=8.

12. how come?

Because 80 minutes equals 1 hour, 20 minutes.

13. shuttle

Ten. As our train leaves Washington, there is a train just 1 hour away from Washington, another 2 hours away, etc. Since the trains are traveling at the same rate of speed, we will pass the train that is 1 hour away from Washington in $\frac{1}{2}$ hour, the train 2 hours away in 1 hour, or $\frac{1}{2}$ hour after passing the preceding train; that is, we pass a train every $\frac{1}{2}$ hour. Since the trip takes $4\frac{1}{2}$ hours, or nine half hours, and as we leave Washington we also pass a train just arriving, we shall pass a total of ten trains during our trip.

14. touching matches

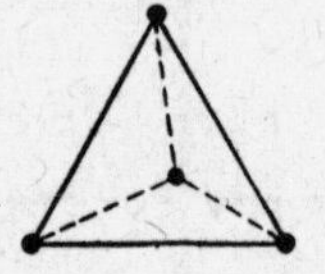

15. get rich quick

The person who bought the pawn ticket, since the pawn ticket is worth only $\$1.00 - \$0.75 = \$0.25$.

16. playing with matches

All Touching

17. gluttonous cats

Six. If six cats eat six rats in 6 minutes, then six cats eat one rat in 1 minute; therefore, six cats would eat a hundred rats in 100 minutes.

18. a lot of bull

1000 pounds.

19. an odd one

Any odd number; for example, $5 \div 2 = 2$ with 1 left over.

20. triangulate

Seven. They form two tetrahedrons with a common base.

21. strike this one out

One half of st(rik)e = st + e; two thirds of t(on) = on; so that, st + e + on = st + on + e = stone.

22. even off

SEVEN − S = EVEN
ELEVEN − EL = EVEN

23. something different

20. Since $2 \times 25 = 50$ and $(2 \times 5) + 20 = 30$, then $50 - 30 = 20$.

24. playing post office

Twelve.

25. six yield three

Remove matches numbered 5, 6, and 9.

26. bye bye blackbirds

The four that were shot. The others flew away. The original number of blackbirds was $(2 \times 4) + 20 = 28$ and the number shot is $\frac{1}{7}$ of 28, or four.

27. too touching

No. For three coins to be at equal distances, all three must be tangent to each other. It is then impossible to place a fourth coin tangent to each of the first three coins.

28. no lost marbles

Yes. Place three marbles on a table tangent to each other, then place the fourth on top of them so that all four marbles are mutually tangent.

29. same but different

$33 - 3$; $3^3 + 3$; $5 \times 5 + 5$; $6 \times 6 - 6$; XXX.

30. two for the summing

$$\begin{array}{r} 11 \\ 1 \\ 1 \\ \underline{1} \\ 14 \end{array}$$

31. it all adds up

$$\begin{array}{r} 1 \\ 2 \\ 34 \\ 56 \\ \underline{7} \\ 100 \end{array}$$

32. figure this one out

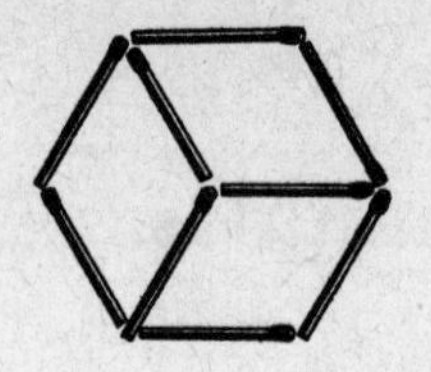

33. double up

A half dollar. Deduct the word half, and you have a dollar.

34. lover boy

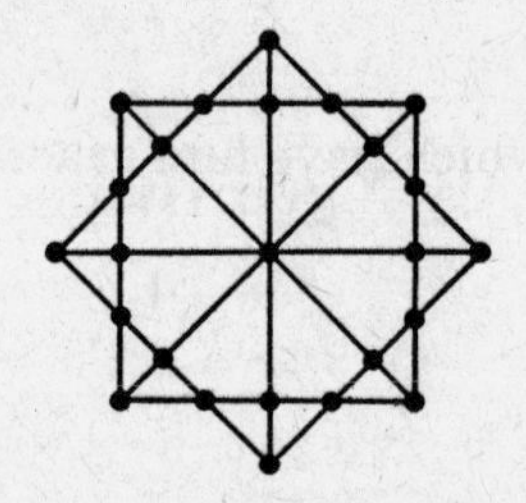

35. lover boy does it again

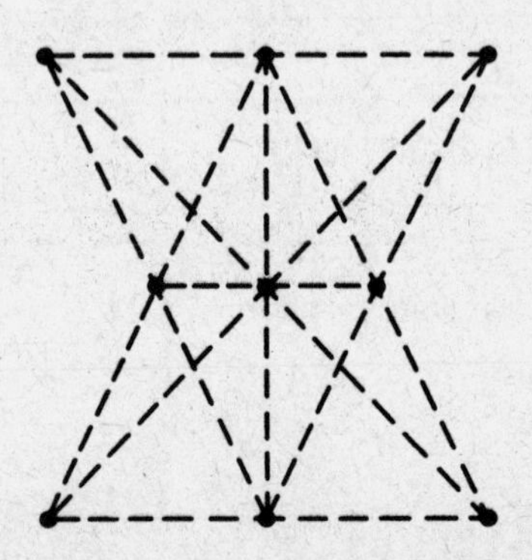

36. weigh carefully

A pound of feathers. Gold is measured in units of troy weight, while feathers are measured in units of avoirdupois weight. Since a troy pound equals 5760 grains and an avoirdupois pound equals 7000 grains, a pound of feathers is 7000 — 5760, or 1240 grains heavier than a pound of gold.

37. water boy

The boy filled the 3 quart jar and poured it into the 5 quart jar. Then he filled the 3 quart jar again and poured it into the 5 quart jar until the 5 quart jar was full. He then had 1 quart left in the 3 quart jar. He emptied the 5 quart jar and poured the 1 quart into it. Finally, he filled the three quart jar and poured it into the 5 quart jar, which gave him exactly 4 quarts.

38. take away and get more

Write 19 = XIX, take away I = 1, and you have XX = 20.

39. equal but not the same

Because a dry quart measure does not equal a liquid quart measure. A liquid quart equals $\frac{1}{4}$ gallon, or 57.75 cubic inches; while a dry quart equals $\frac{1}{8}$ peck, which is 67.20 cubic inches.

40. crack this bottle

The cork is 5¢; the bottle is $1.05. Since the bottle costs $1.00 more than the cork, then twice the cost of the cork plus $1.00 is $1.10; thus, twice the cost of the cork is 10¢. Therefore, the cork costs 5¢ and the bottle costs $1.05.

41. mental exercise

1275. Arrange the numbers in pairs: 1 and 50, 2 and 49, 3 and 48, 4 and 47, etc. The sum of each pair is 51 and, since the total number of pairs is 25, the total sum is $51 \times 25 = 1275$.

42. letter perfect

$\frac{1}{3}$ of TW(EL)VE = EL
$\frac{4}{5}$ of S(EVEN) = EVEN
EL + EVEN = ELEVEN

43. a re-match

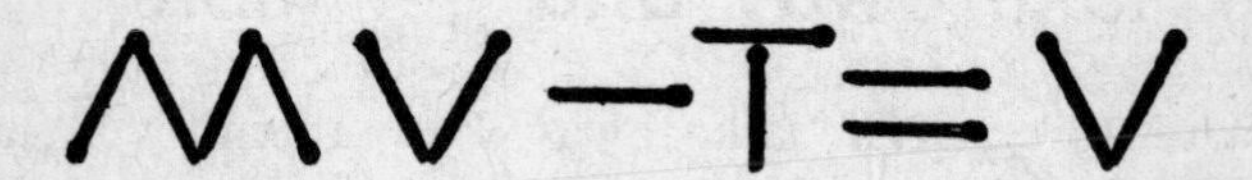

44. nutty arithmetic

Write 4 = IV, turn it upside down IΛ, put IV on top of it and you have IX = 9.

45. a neat profit

$1. If 100% of 1¢ equals 1¢, then 10,000% of 1¢ equals 100×100 % of 1¢, or $100 \times$ 1¢.

46. a-hunting we will go

$403\frac{1}{5}$ rods. The dogs gain $21 - 16 = 5$ rods per unit time. Four score plus sixteen is $(4 \times 20) + 16 = 96$ rods. Hence, the dogs run $96 \div 5 = 19\frac{1}{5}$ units of time. Therefore, the distance the dogs run before catching hare is $19\frac{1}{5} \times 21 = 403\frac{1}{5}$ rods.

47. good neighbor policy

After the oldest brother took his share, $\frac{1}{2}$ of 17, or $\frac{17}{2}$, $\frac{17}{2}$ was left over. From this, the second brother subtracted his share, $\frac{17}{3}$, thus $\frac{17}{2} - \frac{17}{3} = \frac{17}{6}$. From this $\frac{17}{6}$ the third brother deducted his share, $\frac{17}{9}$, thus $\frac{17}{6} - \frac{17}{9} = \frac{17}{18}$. Clearly, $\frac{1}{18}$ was not accounted for in this division of the estate.

48. more nutty arithmetic

Write $9 = \text{IX}$, cross the I with another $\text{I} = 1$, and you have $\text{XX} = 20$.

49. a horse laugh at lots of corn

1 ton. If 100 horses eat 100 tons of corn in 100 days, then 100 horses will eat 1 ton of corn in 1 day, and it follows that 100 horses will eat 10 tons of corn in 10 days, thus 10 horses will eat 1 ton of corn in 10 days.

50. lad and dad

They are the same height. Measure them.

51. apple sauce

15¢. Since one third of six, or two apples, sells for 3¢ and one fourth of forty apples is ten apples, which is five times two apples, ten apples sell for $5 \times 3 = 15$¢.

52. medium to rare

10 pounds. Let W = total weight, then $W = 5W/2$, or $W - W/2 = 5$; hence $2W - W = 10$, or $W = 10$ pounds.

53. duck this one

Three ducks. Let D = the number of ducks, then $3D = 2D + 3$, or $3D - 2D = 3$; hence, $D = 3$.

54. get to the root of this one

$$V - IV = \sqrt{1}$$

(the square root of 1)

55. the jealous husbands

Let H_1, H_2, and H_3 represent the husbands and W_1, W_2, and W_3 represent wives; then this is how they crossed the river:

1. W_1 and W_2 cross, W_2 returns.
2. W_2 and W_3 cross, W_3 returns.
3. H_1 and H_2 cross, H_2 and W_2 return.

4. H_2 and H_3 cross, W_1 returns.
5. W_1 and W_2 cross, W_2 returns.
6. W_2 and W_3 cross.

56. the old and the new

18 years old. Let my age be A, then $5 \times 7 + 7 \times 3 = 35 + 21 = 56$ and $6 \times 9 + 4 = 58$. Thus, the difference between 56 added to my age and 58 equals the difference between twice my age and 20. Therefore, we have this equation,

$$
\begin{aligned}
(A + 56) - 58 &= 2A - 20 \\
A - 2 &= 2A - 20 \\
A - 2A &= -20 + 2 \\
A &= 18.
\end{aligned}
$$

57. all lines are equal

The conclusion is based on the erroneous idea that a straight line is made up of a very large, but finite, number of points. Since a point has only location, but no dimensions, a line, which has length, cannot be constructed from points. The concept of a straight line is accepted as axiomatic, hence no definition of it is necessary. A straight line, however, may be thought of as the path traced by a point moving in a constant direction.

58. the student's blunder

540. If the number divided by 6 equals 15, then the number is equal to $15 \times 6 = 90$, and the correct answer is $90 \times 6 = 540$.

59. a fish story

None. There are no fish in the basket.

60. writer's cramp

$44 + \frac{44}{4}$.

61. ancient write-in

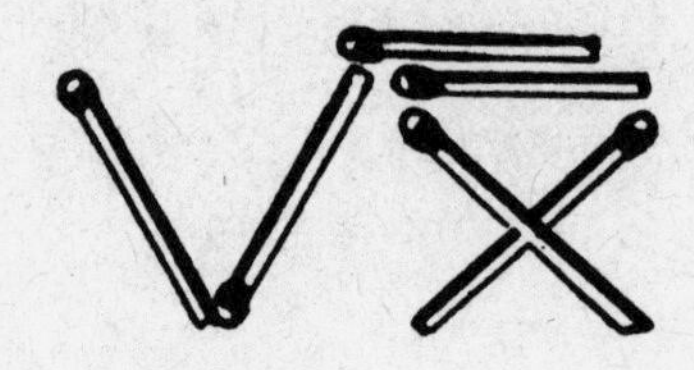

62. sweet in his cups

Put three pieces of sugar into one cup and seven pieces into another cup, then put one of the cups into the third cup.

63. stu simple

For the sake of justifying "Pit Agoras," the operation should be,

$$\begin{array}{r} 4 \\ 7\overline{)28} \\ 28 \\ \hline 0 \end{array}$$

64. nothing to it

Draw a horizontal line through the number 888, dividing it in half; thus, half of ~~888~~ is 000 = 0.

65. too many eggs in one basket

The first farmer was asking $\frac{1}{2}$¢ for an egg; thus, he should have received $\frac{1}{2} \times \$3.00 = \$1.50$. The second farmer was asking $\frac{1}{3}$¢ for an egg; hence, he should have received $\frac{1}{3} \times \$3.00 = \$1.00$. Therefore, both together should have received $\$1.50 + \$1.00 = \$2.50$. However, by combining their eggs, they got only $\frac{2}{5}$¢ for an egg; that is, *the rate was changed,* so that they received $\frac{2}{5} \times \$6.00 = \$2.40$. Hence, together they lost $\$2.50 - \$2.40 = 10$¢; however, the second farmer received $\$1.20 - \$1.00 = 20$¢ more, while the first lost $\$1.50 - \$1.20 = 30$¢.

66. matchless figures

Remove matches numbered 4, 5, 2, and 7.

67. grouping

$$2\,\tfrac{2}{2} + 2 = 2 + 1 + 2 = 5.$$

68. cut it out

Make a circular cut dividing the pie into two equal parts. Then make two straight cuts through the center, dividing the pie into eight equal parts. See teaser No. 15, chapter 3.

69. use a fourth dimension

Push the cork in.

70. the tender missionaries and the hungry cannibals

Let M_1, M_2, and M_3 represent the missionaries and let C_1, C_2, and C_3 represent the cannibals, with C_3 able to operate the outboard motor; then this is how they crossed the river:

1. C_1 and C_3 cross, C_3 returns.
2. C_2 and C_3 cross, C_3 returns.
3. M_1 and M_2 cross, C_1 and M_2 return.
4. M_2 and C_3 cross, M_1 and C_2 return.
5. M_1 and M_3 cross, C_3 returns.
6. C_1 and C_3 cross, C_3 returns.
7. C_2 and C_3 cross.

chapter 3
teasers for the mathematicians

1. the cup that cheers

A recipe for punch calls for a third and a half of a third of four cups of brandy. How many cups of brandy does the recipe require?

2. the land of ifthen

If a quarter of twenty is four, what would a third of ten be?

3. melon fare

A watermelon weighs nine tenths of its weight plus $\frac{9}{10}$ of a pound. If I eat a slice consisting of one ninth of the melon, what is the weight of the slice?

4. a square in ifthen

If three times five were twenty, what would the square of six be?

5. metamorphosis

Transform the fraction into an integer by replacing just one nail.

6. balancing act

If a brick balances with three quarters of a brick of the same weight and size plus three quarters of a pound, how much does the brick weigh?

7. long way around

Write the number 1 using each of the nine digits 1, 2, 3, 4, 5, 6, 7, 8, and 9 only once.

8. in wonderland

$$\begin{array}{r} 3414 \\ 340 \\ 74813 \\ \hline 433748.13 \end{array}$$

Show that this sum is correct.

9. sloppy joe

Sloppy Joe never puts away his socks in pairs, but simply throws them into a drawer. There are fourteen black socks and eight blue socks in the drawer. One night, he needed just one pair of socks the same color, but, in order not to wake up his roommate, Joe did not put the light on; therefore, he had to get the socks in complete darkness. How many socks did Joe have to draw to be sure that he had a pair of matching socks.

10. get a kick from this one

An old wine maker has a cask half full of 3 year old wine and another cask twice its size which is one third full of 3 year old wine. If both casks are filled with 1 year old wine and their contents are mixed, what part of the mixture is 3 year old wine?

11. count them out

Joe in five hours a sum can count,
Which Dick can in eleven;
How much more then is the amount
They both can count in seven?

12. transformation

Transform the expression,

into the number 100 by rearranging the matches.

13. fence 'round

Enclose 12 square inches with twelve matches, each 2 inches long.

14. a wail of woe

Jack complained to Jill that he had agreed to pay $800 in cash plus a fixed number of bushels of wheat as the yearly rental for his farm. That, he explained, would amount to $70 an acre when wheat was worth 75¢ a bushel. Since wheat is now worth $1 a bushel, he must pay $80 an acre, which he thought was too much. What is the size of the farm?

15. half and half

If a pie is divided into two equal parts by a circular cut, what is the radius of the circular cut?

16. this and that

This and that plus a half of this and that is what percent of this and that?

17. lay away plan

If 100 chickens lay 10,000 eggs in 100 days, how many days will it take one chicken to lay one egg?

18. back in the land of ifthen

What would ten be if four were six?

19. landscape gardening

A landscape architect wishes to divide a circular garden into four similar tracts, each tract bounded by arcs. How can he do this?

20. fraction less

Without the use of fractions, express the number eight by using eight eights.

21. yoke mates

A husband is 10 years older than his wife. If the sum of their ages is five times their difference, what are their ages?

22. rhombus anyone

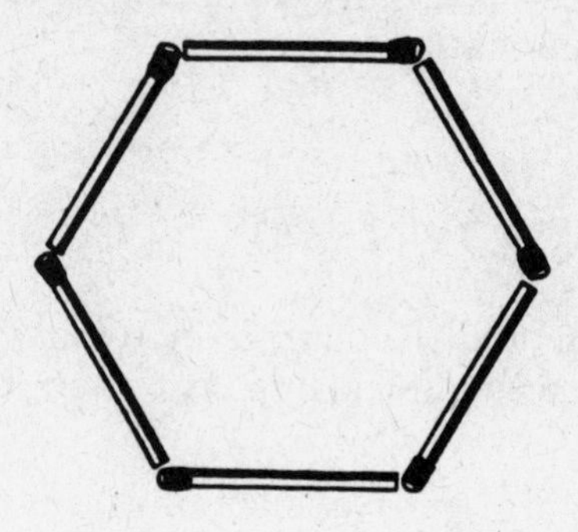

In this arrangement there are six matches forming a regular hexagon. Rearrange two matches and add one more match to form two rhombuses. (A rhombus is an equilateral parallelogram having oblique angles.)

23. nothing to wear

A lady showed her husband five blouses, three skirts, and four pairs of shoes, all of which match, but she complained to him that she had nothing to wear. Her husband then told her that, with these garments, she could wear a different outfit every day of the month. Show that what her husband said is true.

24. ship ahoy

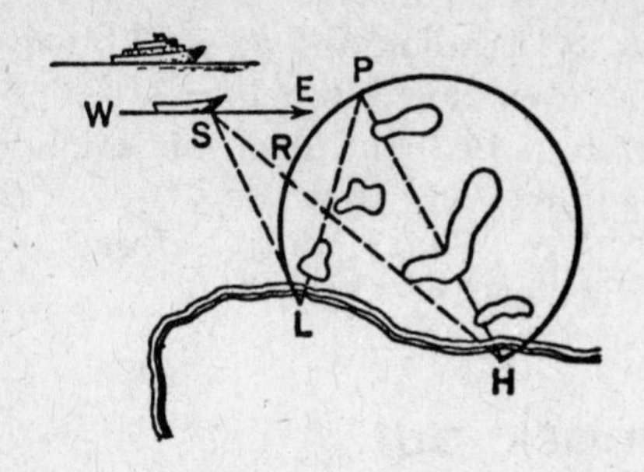

The captain of a ship sailing along an easterly course knows from the chart the location of certain reefs situated off the coast near two lighthouses L and H. The chart also directs his attention to angle LPH, known as the "horizontal danger angle," which is inscribed in a circle passing through L and H and surrounding the reefs. To safeguard his ship from the reefs, the captain must avoid this circle by veering to the left. Which two geometric facts does the captain need to use in selecting his course?

25. cocktail party

If at a cocktail party there are more than 366 persons, why can you be certain that at least two of them have the same birthday?

26. fortune hunters

A man made a will bequeathing one third of his fortune to his wife and two thirds to a son if they should have one, but one half to his wife and the other half to a daughter if they should have a daughter. After his death, twins are born, a son and a daughter. How should the fortune be divided among the three?

27. a good mixer

Paint sells at $3 a gallon and paint thinner at $1 for 3 gallons. A painter gave his helper $10 and two empty cans, telling him to bring back an equal quantity of paint and thinner, and that the $10 was to cover the total cost exactly. What quantity of each item did the painter's helper get?

28. a knock out

The owner of a sports stadium refused an offer of $2,000,000 for the television rights for the month of March, but said that he would take 1¢ for the first day, double that, or 2¢ for the second day, double this amount, or 4¢ for the third day, and so on for every day in the month. Would he get more or less in this transaction?

29. a stick out

A merchant makes 20% instead of 40% on goods, because his "yardstick" is too long. What is the actual length of the yardstick?

30. more of this and that

This and that plus two thirds of this and that is what part of four thirds of this and that?

31. a halloween tale

At a country fair Hank bought a prize pumpkin that weighed 10 pounds plus half of 10 pounds and half of its own weight besides. What did the pumpkin weigh?

32. 'riting and 'rithmetic

Transform the expression,

LXVIII = 68

into ten without adding or removing any matches.

33. ass grazing

In the midst of a meadow, well covered with grass,
Just an acre was needed to tether an ass;
How long was the line that reaching all around,
Restricted its grazing to an acre of ground?

34. man 'tis the greatest

What is the largest number you can express with three matches?

35. speedy

A man must travel a distance of 4 miles. If he travels the first 2 miles at 30 miles per hour, at what speed must he travel the remaining 2 miles so that he can average 60 miles per hour in going 4 miles?

36. fore

Three men were about to start a game of golf, but one of them discovered that he had forgotten to bring any golf balls. The first golfer produced five new balls and the second four new balls. They divided the golf balls equally among themselves and the forgetful golfer had to pay $3 for his share of the golf balls. How was the $3 divided equitably between the other two golfers?

37. they are off and running

If a circular track is 5 yards wide and it takes a horse, traveling his fastest, π more seconds to travel the outer edge than the inner edge, what is his rate per second?

38. tailor's apprentice

If a tailor's apprentice, after cutting off 10% of a piece of cloth, had 90 yards left, how many yards did he have at first?

39. par for the duffer

On being asked his golf score, Ray said to Everett, "If I had as many more strokes, plus half as many more, plus fifty to boot, I should have 300." What was Ray's score?

40. back to ifthen

If six is a third of twelve, what would a fourth of twenty be?

41. joe's grandpa

Joe's grandfather has lived a quarter of his life as a boy, a sixth of his life as a young man, half of his life as a middle-aged man, and 6 years as a senior citizen. How old is Joe's grandfather?

42. mow it down

Following the boundary of a rectangular field, it requires 11 rounds of a lawn mower to cut one half of it and 14 more to cut the remainder. If the lawn mower cuts a swath 2 feet wide and all the swaths are full, what are the dimensions of the field?

43. stormy weather

I place a bowl into the storm
To catch the drops of rain;
A half a globe was just its form,
Two feet across the same.
The storm was o'er, the tempest past,
I to the bowl repaired;
Six inches deep the water stood,
It being measured fair.
Suppose a cylinder, whose base
Two feet across within,
Had stood exactly in that place,
What would the depth have been?

44. a lot of nothing

If you had as many more dollars plus twice as many more dollars plus one half as many dollars less $9, you would have nothing. How much do you have?

45. plane fact

Express the number one using four matches lying in the same plane so that no two matches are parallel or lying in the same straight line.

46. a neat profit

What percent of the cost do I gain if I sell five eighths of a bill of goods for what three fourths of what it cost me?

47. partners

Dick and Bob form a partnership in a store. Dick furnishes the money and half the rent, while Bob works in the store and furnishes the other half of the rent. Bob being short of money one day, uses $60 from Dick's money to pay the rent. The next day they receive from a customer an amount of money in which Dick owns a third interest. How many dollars should be returned to Dick from this amount to maintain his equity?

48. dog gone

A dog is at the center of a circular pond 200 feet in diameter and a duck is swimming around the outer edge of the pond. The dog starts toward the duck swimming at the same speed as the duck. If the dog continually keeps in line with the center of the pond and the duck, how far must he swim before reaching the duck?

49. average rate

A man travels to Austin, Texas at 40 miles per hour and returns at 60 miles per hour. What was his average speed?

50. trip along

If a man traveling a certain distance increases his speed by one fifth, so that he makes the trip in 6 hours, how long does it take him to make the same trip going at his original speed?

51. vat is loss?

It takes one day to fill the vat
With this large pipe, two days with that;

The third pipe needs but one day more;
The fourth pipe fills the vat in four.
If all four pipes together run,
How long before the task is done?

52. middle spread

A plane perpendicular to the earth's axis and halfway between its poles cuts through the earth's surface at the equator, which describes a circle on this plane. Every point on this circle is at the same distance from the north and south poles. If 48 inches are added to the earth's equator, what change is made in the earth's radius?

53. the short and the long way

If 1 degree at the equator is approximately 69.1 miles, and on a circle of latitude passing through New York City, 1 degree is approximately 52.3 miles, how many more miles would a person making a trip around the equator have to travel than a person making a trip around the latitude circle passing through New York City?

54. double in ifthen

If two thirds of a number is added to twenty-four, the number will be doubled. What is the number?

55. the mini number

Find the smallest number which divided by each of the integers 2, 3, 4, 5, 6, 7, 8, 9, and 10 will give, in each case, a remainder which is 1 less than the divisor.

56. ships coming in

Three ships leave New York for Le Havre, France, on the same day. It takes the first ship 12 days, the second ship 16 days, and the third ship 20 days to make a round trip. How many days will elapse before all three ships again leave New York on the same day, and how many trips will each ship have made in the meantime?

57. the wise farmer

Paul owns a farm situated between two rivers, the West and East rivers. He must deliver produce from point

A to the West River for shipment and then pick up merchandise at the East River for delivery to point B. At which points on these rivers should Paul build piers, so that, in following his usual route, he will travel the shortest distance?

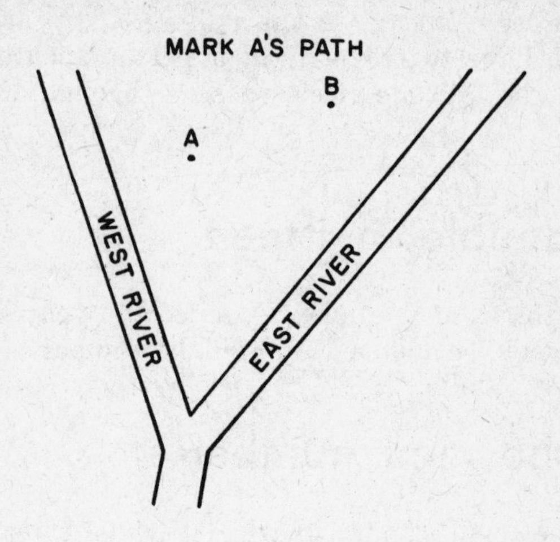

58. a train of thought

A man on a railroad platform observed that a train passed by the point where he was standing in 10 seconds, and that the same train passed completely through a station, which is 308 yards long, in 24 seconds. How long was the train and how fast was it going?

59. passing trains

Two trains pass each other in 10 seconds when moving in opposite directions; when moving in the same direction the swifter train passes the slower one in 25

circumference. If the weight of the segment is 5 pounds, what is the weight of the cheese?

67. said the spider to the fly

A room is 30 feet long, 12 feet wide, and 12 feet high. A spider on the center line of the west wall of the room, one foot above the floor, sees a fly asleep on the center line of the east wall, one foot below the ceiling. The spider wants to get to the fly as soon as possible. Which is the shortest path for the spider to take to get to the fly and what is the length of this path?

68. a walking race

Tony, Chris, and Paul walk $3\frac{1}{2}$, 4, and 5 miles an hour, respectively. They walk on circular tracks which are $\frac{1}{5}$, $\frac{1}{4}$, and $\frac{3}{8}$ of a mile in circumference (in the same order) with their centers on the same straight line. Tony, Chris, and Paul start from points on this line at the same instant.

a) How long will it take before all three boys are again back on the line at the the same time?

b) How long will it take before all three boys are again at the points from which they originally started?

69. "time discovers truth" (Seneca)

A man leaving his office for a business appointment noted the positions of the hands of a clock. Between 2 and 3 hours later he came back and found that the hands had exchanged places. How long was he away from the office?

70. "the stuff that life is made of" (Benjamin Franklin)

A business man looks at his watch before leaving the office for lunch. When he returns, he finds that the hour and minute hands have exchanged places from the positions they had when he left the office. a) Find the time when he left and b) the time when he returned.

71. traffic jam

A certain highway was being repaired, so it was necessary for the traffic to use a detour. At a certain time, a car and a truck met in this detour which was so narrow that neither the truck nor the car was able to pass. Now, the car had gone three times as far into the detour route as the truck had gone, but the truck would take three times as long to reach the point where the car was. If both the car and the truck can move backward at one third of their forward speed, which of these two vehicles should back up in order to permit both to travel through the detour in the minimum amount of time?

72. homework

A teacher assigned homework to his class and told the students that on each day after the first, they must do twice the number of problems that they had done so far. If at the end of five days, the students had completed one third of the problems, how long will it take them to do all of their problems?

solutions

1. the cup that cheers

2 cups. Since a third of 4 cups is $\frac{4}{3}$ cups, and a half of a third of 4 cups is $\frac{1}{2} \times \frac{4}{3} = \frac{2}{3}$ of a cup, a third and a half of a third of 4 cups is $\frac{4}{3} + \frac{2}{3} = \frac{6}{3} = 2$ cups.

2. the land of ifthen

$2\frac{2}{3}$. If $\frac{1}{4}$ of $20 = 4$, then $5 = 4$ and $10 = 2 \times 5$ becomes $2 \times 4 = 8$; so that $\frac{1}{3}$ of 10 becomes $\frac{1}{3}$ of $8 = 2\frac{2}{3}$.

3. melon fare

1 pound. If $\frac{9}{10}$ of its weight plus $\frac{9}{10}$ of a pound equals the weight of the melon, then $\frac{9}{10}$ of a pound is $\frac{1}{10}$ the weight of the melon; hence, 9 pounds is the weight of the melon. Thence, it follows that the weight of the slice is $\frac{1}{9}$ of the melon, or 1 pound.

4. a square in ifthen

48. If $3 \times 5 = 20$, then $3 = 4$; hence $6 = 2 \times 3$ becomes $2 \times 4 = 8$ and the square of $6 = 6 \times 6$ becomes $6 \times 8 = 48$. Also, if $3 \times 5 = 20$ and $6 \times 6 = x$, then $15:36 = 20:x$ and $x = 48$.

5. metamorphosis

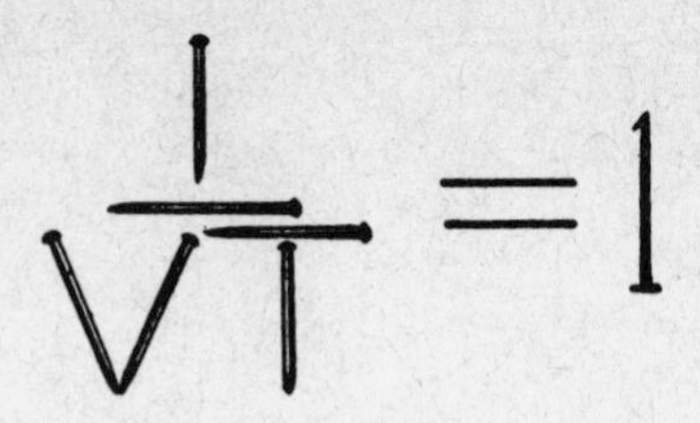

6. balancing act

3 pounds. If $\frac{3}{4}$ the weight of a brick plus $\frac{3}{4}$ of a pound equals the weight of the brick, then $\frac{3}{4}$ of a pound is $\frac{1}{4}$ the weight of the brick; therefore, the weight of the brick is 3 pounds.

7. long way around

$1^{23456789}$.

8. in wonderland

Hold the page in front of a mirror and it will read,

NINE

ONE

EIGHT

EIGHTEEN

9. sloppy joe

Three socks. If the first sock drawn is blue and the second is black, then the third must match either one of the first two, since it must be either blue or black.

10. get a kick from this one

$\frac{7}{18}$. A cask half full of 3 year old wine is $\frac{1}{4}$ of a cask twice its size. Therefore, $\frac{1}{4} + \frac{1}{3} = \frac{3+4}{12} = \frac{7}{12}$ of the large cask contains 3 year old wine. After filling both casks with 1 year old wine, the small cask is half full of 1 year old wine and the large cask is $\frac{2}{3}$ full of 1 year old wine. Therefore, $\frac{1}{4} + \frac{2}{3} = \frac{3+8}{12} = \frac{11}{12}$ of the large cask contains 1 year old wine. Since $\frac{7}{12}$ of the large cask contains 3 year old wine and $\frac{11}{12}$ of it contains 1 year old wine, it follows that the total amount of wine, both new and old, is $\frac{18}{12}$. Thus, $\frac{7}{12} \div \frac{18}{12} = \frac{7}{12} \times \frac{12}{18} = \frac{84}{216} = \frac{7}{18}$. Therefore, $\frac{7}{18}$ of the mixture is 3 year old wine.

11. count them out

$1\frac{2}{55}$. In 1 hour Joe can count $\frac{1}{5}$ of the sum and in 1 hour Dick can count $\frac{1}{11}$ of it; therefore, in 1 hour both can count $\frac{1}{5} + \frac{1}{11} = \frac{11+5}{55} = \frac{16}{55}$ of the sum, and in 7 hours both can count $7 \times \frac{16}{55} = \frac{112}{55} = 2\frac{2}{55}$ sums. Therefore, both can count $2\frac{2}{55} - 1 = 1\frac{2}{55}$ sums more.

12. transformation

$$\sqrt{\overline{X}} = \sqrt{10{,}000} = 100$$

13. fence 'round

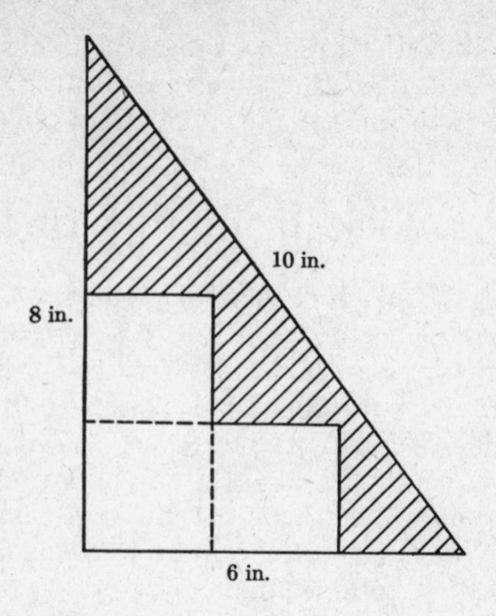

The shaded area in the right triangle is bounded by the twelve matches. The area of the triangle is one half its base times the height, or 24 square inches. The area of the three unshaded squares is, therefore, 12 square inches. Hence, the shaded area is $24 - 12 = 12$ square inches.

14. a wail of woe

20 acres. A difference of $\$1.00 - \$0.75 = \$0.25$ in the price of wheat per bushel makes a difference of \$10 an acre in rent. Hence, the rent paid in wheat is $\$10 \div \$0.25 = 40$ bushels per acre. The value of 40 bushels of wheat then is $40 \times \$1 = \$40$, so that the rent per acre paid in cash is $\$80 - \$40 = \$40$ per acre. Therefore, the number of acres is $\$800 \div \$40 = 20$ acres.

15. half and half

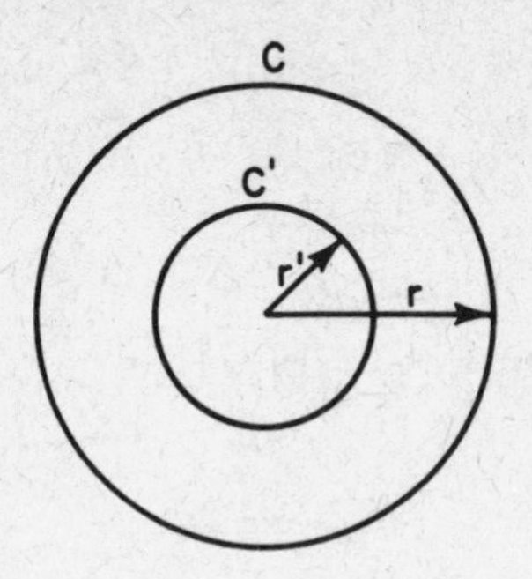

Approximately $\frac{7}{10}$ the radius of the pie. Since the pie is circular, its area is $A = \pi r^2$, where A is the area and r is the radius. Letting the area and radius of the inner circular cut be A' and r', respectively, then $A = 2A'$, hence $\pi r^2 = 2\pi r'^2$ and $r' = \frac{r}{\sqrt{2}}$, which is approximately $\frac{7}{10}r$. (See also problem 68, chapter 2).

16. this and that

150%. This and that and a half of this and that is $\frac{3}{2}$ of this and that, which is one and a half times, or 150% of this and that.

17. lay away plan

1 day. If 100 chickens lay 10,000 eggs in 100 days, then 100 chickens can lay 100 eggs in one day, and one chicken can lay one egg in 1 day.

18. back in the land of ifthen

15. If x equals what ten would be, then $10:x = 4:6$, and $x = 15$.

19. landscape gardening

20. fraction less

$8(88888)^{8-8} = 8(88888)^0 = 8(1) = 8.$

21. yoke mates

The sum is $5 \times 10 = 50$ years.
The wife's age is $\frac{1}{2}(50 - 10) = 20$ years.
The husband's age is $20 + 10 = 30$ years.

22. rhombus anyone

The dotted lines show the original positions of the two matches.

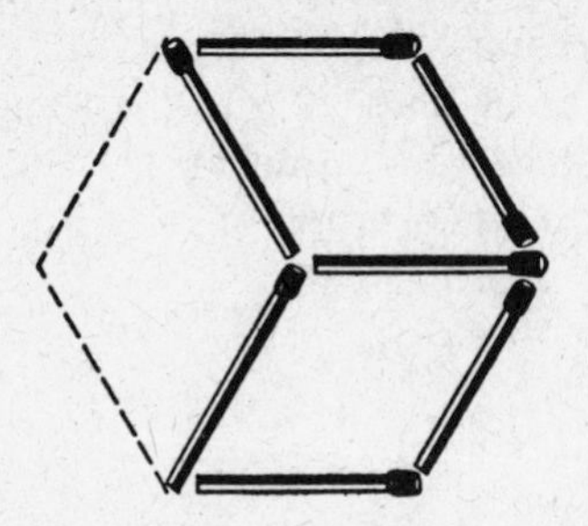

23. nothing to wear

With the selection of any one of five blouses there may be combined any one of three skirts. Thus, there are $3 \times 5 = 15$ different ways of selecting a blouse and skirt combination. For each of these selections, in turn, there are four possible ways of selecting a pair of shoes. Therefore, there are a total of $4 \times 15 = 60$ possible outfits to choose from.

24. ship ahoy

First, an angle inscribed in a circle is measured by one half its intercepted arc. Thus, angle LPH = $\frac{1}{2}$ arc LH. Second, an angle formed by two secants intersecting outside a circle is measured by one half the difference between the intercepted arcs. Thus, angle LSH = $\frac{1}{2}$ (arc LH − arc RL) = $\frac{1}{2}$ arc LH − $\frac{1}{2}$ arc RL. Therefore, angle LSH is less than angle LPH. As long as the captain makes sure that angle LSH is less than horizontal danger angle LPH, the ship is outside the circle and is therefore safe from the reefs.

25. cocktail party

If all 366 persons have different birthdays (including February 29 in a leap year), then the 367th person's birthday must be the same as that of at least one of the others, since there are only 366 possible days in a year to choose from.

26. fortune hunters

The wife and the daughter were to receive equal shares, while the son was to receive twice as much as the wife. Hence, if the estate is divided into four equal parts, the son would get two fourths, or one half, while the wife and daughter would each get one fourth.

27. a good mixer

3 gallons. If a gallon of paint costs \$3 and a gallon of paint thinner costs a third of a dollar, then a gallon of each costs $3 + \frac{1}{3} = \$3\frac{1}{3}$. Therefore, the number of gallons of each equals $10 \div 3\frac{1}{3} = 10 \times \frac{3}{10} = 3$ gallons.

28. a knock out

Thus, he would gain \$21,474,836.47 — \$2,000,000.00 = \$19,474,836.47, since the amount he would receive for the month of March is $2^{31} - 1$.

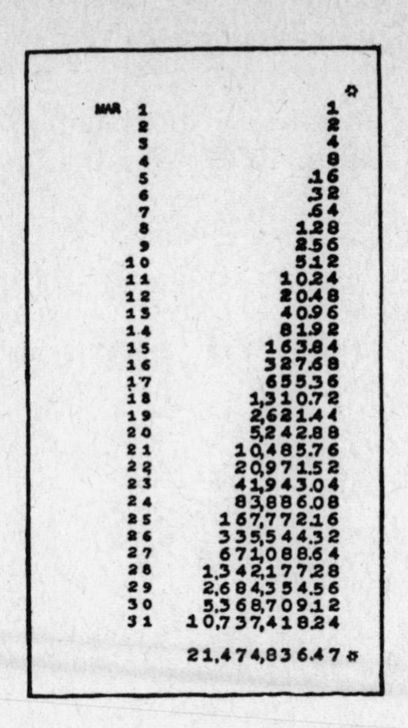

29. a stick out

42 inches. In order to make 40% of cost, the merchant should sell for 140% of cost. But he makes 20% of cost, so that he actually sells for only 120% of cost. Since a yard is 36 inches, then 120% of the actual length is 140% of 36, so that the actual length of the stick is $\frac{140\%}{120\%} \times 36 = \frac{7}{6} \times 36 = 42$ inches.

30. more of this and that

$1\frac{1}{4}$ of $\frac{4}{3}$. If this and that plus $\frac{2}{3}$ of this and that is $1 + \frac{2}{3} = \frac{5}{3}$ of this and that, it follows that $\frac{5}{3}$ of this and that is $\frac{5}{3} \div \frac{4}{3} = \frac{5}{3} \times \frac{3}{4} = \frac{5}{4}$; hence, $\frac{4}{3} = 1\frac{1}{4}$ of this and that.

31. a halloween tale

30 pounds. The weight W of the pumpkin equals $10 + \frac{1}{2}(10) + \frac{1}{2}W$. Hence, $15 = \frac{1}{2}W$; therefore, $W = 30$ pounds.

32. 'riting and 'rithmetic

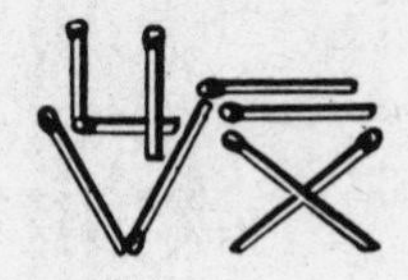

33. ass grazing

39.24 yards. Let r be the length of the line; the area grazed over is 1 acre = 4840 square yards. Hence, $\pi r^2 = 4840$; $r^2 = 4840 \div \pi$. Assume $\pi = \frac{22}{7}$ (approx.), then $r^2 = 4840 \div \frac{22}{7} = 1540$ and $r = \sqrt{1540} = 39.24$ yards (approx.).

34. man 'tis the greatest

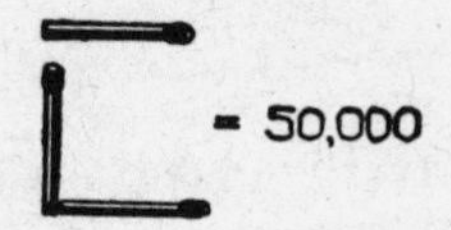

35. speedy

None. To average 60 miles per hour for 4 miles, the man must travel the 4 miles in $\frac{4}{60} = \frac{1}{15}$ of an hour. But in traveling the first 2 miles at 30 miles per hour he has already used $\frac{2}{30}$, or $\frac{1}{15}$ of an hour, so that he has no time left; therefore, this cannot be done.

36. fore

Since the first golfer contributed two golf balls and the second contributed one golf ball. For the distribution to be equitable, the first golfer received $\frac{2}{3}$ of the $3, or $2, while the second received $\frac{1}{3}$ of the $3, or $1.

37. they are off and running

10 yards per second. Since width of track is 5 yards, if r is radius of inner edge of circular track, then $2\pi(r + 5)$ is the distance traveled on the outer edge and $2\pi r$ is the distance traveled on the inner edge. Hence, horse travels $2\pi(r + 5) - 2\pi r = 10\pi$ yards farther on the outer edge than on the inner edge in π seconds. Therefore, in 1 second he would travel $10\pi \div \pi = 10$ yards; that is, his rate is 10 yards per second.

38. tailor's apprentice

100 yards. After cutting off 10%, the tailor's apprentice had $100\% - 10\% = 90\%$ of the piece of cloth remaining. If 90% of the piece of cloth is 90 yards, then 1% of the piece of cloth is 1 yard and the piece of cloth is 100 yards.

39. par for the duffer

100. $2\frac{1}{2}$ times the number of strokes plus 50 is 300, so that $2\frac{1}{2}$ times the number of strokes equals 250; therefore, number of strokes is $250 \div 2\frac{1}{2} = 250 \times \frac{2}{5} = 100$.

40. back to ifthen

$7\frac{1}{2}$. If 6 is $\frac{1}{3} \times 12$, then 6 becomes 4. Since $\frac{1}{4}$ of 20 is 5 and $5 = \frac{5}{4}$ of 4, then $\frac{1}{4}$ of 20 is $\frac{5}{4}$ of $6 = \frac{5}{4} \times 6 = \frac{30}{4} = 7\frac{1}{2}$.

41. joe's grandpa

72 years old. Since $\frac{1}{4} + \frac{1}{6} + \frac{1}{2} = \frac{3+2+6}{12} = \frac{11}{12}$, then $\frac{11}{12}$ of the grandfather's age plus 6 years equals his age. It then follows that 6 years is $\frac{1}{12}$ of his age, so that his age is $6 \times 12 = 72$.

42. mow it down

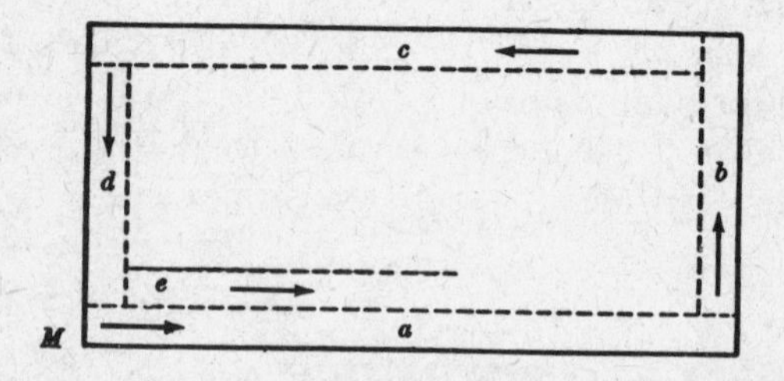

$100 \times 410\frac{2}{3}$ ft. The diagram represents the lawn. The man starts the lawn mower at *M* and cuts the strip *a* down the entire length of the field. Next, he cuts *b*, then *c* and *d*, and is now ready to start on strip *e*. Although the strips *a*, *b*, *c*, and *d* are all of different lengths, he has cut an area which is equal to the difference in area of two rectangles. Since it requires $11 + 14 = 25$ rounds to cut the field and each round consists of two swaths, each 2 feet wide, then the width of the field is $25 \times (2 \times 2) = 100$ feet. The 11 rounds represent 22 swaths of 2 feet each, or $22 \times 2 = 44$ feet. Hence, after 11 rounds the width of the area remaining is $100 - 44 = 56$ feet. Now, if L represents the length of the field, then $(L - 44)(56)$ equals the area remaining. But the area remaining is $\frac{1}{2}$ the total area which equals $\frac{1}{2}(100)L = 50L$. Therefore, $(L - (44)(56) = 50L$, or $6L = 2464$ and $L = \frac{2464}{6} = 410\frac{2}{3}$ feet. Thus, the dimensions of the field are $100 \times 410\frac{2}{3}$ feet.

43. stormy weather

$2\frac{1}{2}$ inches. The amount of water is the volume of a spherical segment of one base with radius NS and altitude PN. The formula for the volume of a spherical segment of one base is $V = \frac{1}{3}\pi h^2(3r - h)$, where r is the radius of the sphere and h is the altitude of the segment. Here $r = 12$ inches and $h = 6$ inches, so that $V = \frac{1}{3}\pi(36)[3(12) - 6] = 360\pi$. Since the volume of a cylinder is $V = \pi r^2 h$, where the radius $r = 12$ inches, then $\pi(144)h = 360\pi$; therefore, $h = \frac{360}{144} = 2\frac{1}{2}$ inches.

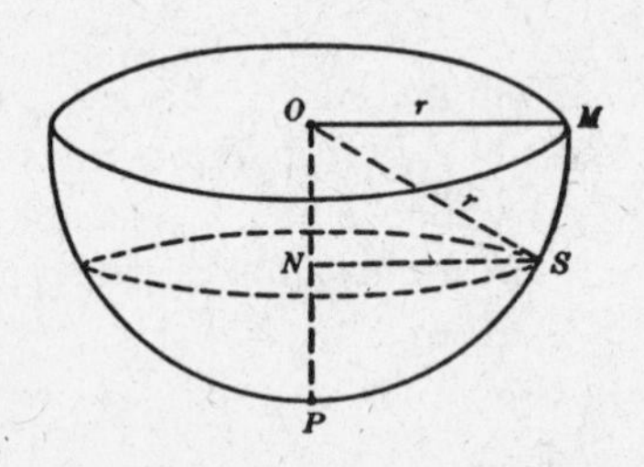

44. a lot of nothing

\$2.00. As many more dollars, twice as many more, and one half as many is equal to $4\frac{1}{2}$ times the number of dollars. Thus, $4\frac{1}{2}$ times the number of dollars is \$9.00 and, therefore, the number of dollars is $\$9.00 \div 4\frac{1}{2} = 9 \times \frac{2}{9} = \$2.00$.

45. plane fact

46. a neat profit

20%. If $\frac{5}{8}$ of the goods are sold for $\frac{3}{4}$ of the cost, then $\frac{1}{8}$ of the goods are sold for $\frac{3}{4} \div 5 = \frac{3}{20}$ of the cost; hence, the goods are sold for $\frac{3}{20} \times 8 = \frac{24}{20} = \frac{6}{5}$, or 120% of cost. Therefore, the profit is 20% of the cost.

47. partners

\$45. Since Dick furnishes half the rent, Bob owes him \$30; but, since Dick owns a third interest, he should receive an amount which exceeds \$30 by a third. Thus, \$30 is $\frac{2}{3}$ of the amount he should receive, so that $\frac{2x}{3} = \$30$, or $x = \$45$.

48. dog gone

Approximately 157.1 feet. The dog follows arc DEB which is half the circumference of a circle whose diameter is 100 feet; therefore, he swims a distance of $\frac{1}{2}(100\pi) = 50\pi$. Letting $\pi = \frac{22}{7}$ (approx), then the distance is $50 \times \frac{22}{7} = 157.1$ feet (approx).

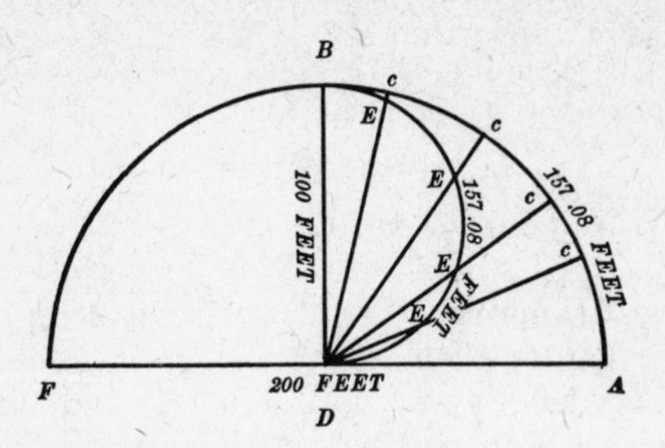

49. average rate

48 miles per hour. Going to Austin, the man travels 1 mile in $\frac{1}{40}$ of an hour and returning, he travels 1 mile in $\frac{1}{60}$ of an hour; hence, for the round trip he averages 1 mile in $\frac{1}{2}(\frac{1}{40} + \frac{1}{60}) = \frac{1}{2}(\frac{3+2}{120}) = \frac{1}{2} \times \frac{1}{24} = \frac{1}{48}$ of an hour. Therefore, his average speed is 48 miles per hour.

50. trip along

$7\frac{1}{5}$ hours. Traveling at $\frac{1}{5}$ of his speed, it will take him 6 times as long, or 36 hours; therefore, going at his regular speed, which is five times faster than $\frac{1}{5}$ of his rate, it will take him $\frac{1}{5}$ of 36 hours, or $7\frac{1}{5}$ hours.

51. vat is loss?

$\frac{12}{25}$ days. The first, or large, pipe will fill $\frac{1}{24}$ of the vat in 1 hour, the second will fill $\frac{1}{48}$ of the vat in 1 hour, the third will fill $\frac{1}{72}$ of the vat in 1 hour, and the fourth will fill $\frac{1}{96}$ of the vat in 1 hour. Hence, all pipes together will fill $\frac{1}{24} + \frac{1}{48} + \frac{1}{72} + \frac{1}{96} = \frac{12+6+4+3}{288} = \frac{25}{288}$ of the vat in 1 hour. It follows that altogether they will fill $\frac{1}{288}$ of the vat in $\frac{1}{25}$ of an hour and the whole vat in $\frac{288}{25}$ hours. Therefore, it will take the four pipes running at the same time; thus $\frac{288}{25} \div 24 = \frac{288}{600} = \frac{12}{25}$ day, or slightly less than $\frac{1}{2}$ day.

52. middle spread

7.6 inches (approx). Let $2\pi r$ be the length of the earth's equator; then, if 48 inches is added to the equator, its length will be $2\pi r + 48$, and the earth's radius r will be increased by x inches, so that a new circle will be formed with radius $r + x$. Constructing a *proportion* between both circles, it then follows that,

$$r : (r + x) = 2\pi r : (2\pi r + 48), \text{ or}$$

$$\frac{r}{r + x} = \frac{2\pi r}{2\pi r + 48}.$$

Therefore, $x = \frac{24}{\pi}$. Letting $\pi = \frac{22}{7}$ (approx), then $x = 24 \times \frac{7}{22} = 7.6$ inches (approx).

53. the short and the long way

6048 miles. Since there are 360 degrees in a circle, then the length of the equator is 69.1 miles $\times$ 360° $=$ 24,876 miles and the length of New York City's latitude is 53.2 miles $\times$ 360° $=$ 18,828 miles; hence the difference, $24{,}876 - 18{,}828 = 6048$ miles.

54. double in ifthen

18. Two thirds of the number plus 24 is two times the number. If $\frac{2}{3}$ of the number is taken away from two times the number, $\frac{4}{3}$ of the number is left. Hence, $\frac{4}{3}$ of

the number is 24 and, therefore, the number is $24 \div \frac{4}{3} = 24 \times \frac{3}{4} = 18$.

55. the mini number

2519. The smallest number which is divisible by each of the given integers and will give a remainder of zero is the smallest number which contains all the given numbers as factors; that is, the least common multiple. The least common multiple of 2, 3, 4, 5, 6, 7, 8, 9, and 10 is 2520. Therefore, the number is $2520 - 1 = 2519$.

56. ships coming in

The number of days that will elapse before the ships leave New York on the same day is the smallest number that contains all the numbers, 12, 16, and 20 as factors; that is, the least common multiple. Since the least common multiple of 12, 16 and 20 is 240, the number of days that will elapse before the ships leave New York on the same day again is 240. Consequently, the first ship makes $240 \div 12 = 20$ trips, the second ship makes $240 \div 16 = 15$ trips, and the third ship makes $240 \div 20 = 12$ trips.

57. the wise farmer

Draw a line from point A perpendicular to the nearest point on the bank of the West River, and denote this point by S, then extend line AS exactly twice its length to point A′ so that lines AS and A′S are equal. In a similar manner, construct a perpendicular line from point B to point R on the nearest bank of the East River and extend this line twice its length to point B′

so that lines BR and B′R are equal. Draw a line connecting A′ and B′, and denote its point of intersection with the bank of the West River containing point S as P, and its point of intersection with the bank of the East River containing point R as P′. Then the line APP′B equals line A′B′ and is, therefore, the shortest distance for Paul's route.

58. a train of thought

220 yards; 45 miles per hour. To pass completely through the station, the train must travel a distance equal to the length of the station plus its own length. To pass the man in 10 seconds, the train must travel a distance equal to its own length in that time. Hence, $24 - 10$, or 14 seconds, is the time it takes the train to travel 308 yards. From this it follows that the distance traveled by the train in 1 second is $308 \div 14 = 22$ yards. Therefore, the length of the train must be $22 \times 10 = 220$ yards, and, since there are 1760 yards in a mile, the train's speed is $(60 \times 60 \times 22) \div 1760 = 45$ miles per hour.

59. passing trains

$24\frac{6}{11}$ miles per hour. The relative speed of the two trains when moving in opposite directions is $(500 + 300) \div 10 = 80$ feet per second. Their relative speeds when moving in the same direction are $(500 - 300) \div 25 = 8$ feet per second. Therefore, the speed of the faster than is $\frac{1}{2}(80 + 8) = 44$ feet per second, or $(44 \times 60 \times 60) \div 5280 = 30$ miles per hour, and the speed of the slower is $\frac{1}{2}(80 - 8) = 36$ feet per second. or $(36 \times 60 \times 60) \div 5280 = 24\frac{6}{11}$ miles per hour.

60. traveling salesmen

42 days. The number of days that will elapse before the salesmen leave the office on the same day is the smallest number that contains all of the numbers 7, 14, and 21 as factors; that is, the least common multiple of these numbers, which is 42 days.

61. tinker's boast

10 pounds. For the pieces to be of equal weight and for the weight to be as great as possible, the greatest number which is a factor common to all the numbers 650, 680, and 760 must be found; that is, the greatest common divisor of these numbers, which is 10.

62. cheese cakes

7.14 inches (approx). Let $5x$ = radius of the top, then $3x$ = radius of the bottom. The kettle is a frustum of a cone and its volume is $V = \frac{1}{3}h(\mathrm{B} + \mathrm{S} + \sqrt{\mathrm{BS}})$, where h is the altitude, S the area of upper base, and B the area of lower base. Here, $h = 12$, $\mathrm{B} = 9\pi x^2$ and $\mathrm{S} = 25\pi x^2$, so that $\mathrm{BS} = 225\pi^2 x^4$ and $\sqrt{\mathrm{BS}} = 15\pi x^2$. Hence,

$$V = \tfrac{1}{3}(12)\,(9\pi x^2 + 25\pi x^2 + 15\pi x^2) = 196\pi x^2,$$

Since seven less than a score of gallons is 13 gallons and 1 gallon is 231 cubic inches, $V = 13 \times 231 = 3003$ cubic inches. Hence, it follows that $196\pi x^2 = 3003$. Letting $\pi = \frac{22}{7}$ (approx), then $x = 2.21$. Therefore, the radius of top is $5x = 5 \times 2.21 = 11.05$ inches, and the radius of bottom is $3x = 3 \times 2.38 = 7.14$ inches (approx).

63. divide and succeed

The sum of the subtrahends and the remainder is equal to the dividend; hence, the dividend is $690 + 2415 + 2070 + 1 = 95221$. The divisor must be the greatest number which is a factor common to all the subtrahends; that is, the greatest common divisor of 690, 2415, and 2070, which is 345. Therefore, the divisor is 345 and the quotient 276 can now be found by actual division. Thus,

```
      276
345)95221
    690
    ----
    2622
    2415
    -----
     2071
     2070
     ----
        1
```

64. time on our hands

$\frac{60}{143}$ minute gained each hour. At 12 o'clock, the minute and hour hands are together. One hour, or 60 minutes, later the minute hand is again at 12:00, but the hour hand is at 1:00, which is 5 minutes farther on. Thus, for the minute hand to catch the hour hand, it must travel from 12:00 to 1:00, or 5 minutes, plus any additional time traveled by the hour hand from 1:00 due to the minute hand's movement. Since the distance traveled by the minute hand is 12 times that traveled by the hour hand, the time traveled by the minute hand from 12:00 is equal to 12 times that traveled by the hour hand from 1:00. But the time traveled by the minute hand from 12:00 is equal to the time traveled by the hour hand from 1:00 plus 5 minutes. Hence, 12 times the time traveled by the hour hand from 1:00 is equal to the time traveled by the hour hand from 1:00 plus 5 minutes. It follows that 11 times the time traveled by the hour hand from 1:00 is 5 minutes; hence, the time traveled by the hour hand from 1:00 is $\frac{5}{11}$ of a minute. Thus, if the watch keeps correct time, the hands should be together every $65\frac{5}{11}$ minutes. Therefore, the watch gains $\frac{5}{11}$ of a minute every 65 minutes, so that it gains $\frac{5}{11} \div 65 = \frac{5}{11} \times \frac{1}{65} = \frac{1}{143}$ of a minute every minute, so that it gains $\frac{1}{143} \times 60 = \frac{60}{143}$ of a minute every hour.

65. square peg in round hole

$4\sqrt{2}$ inches. The diagonals of a square are perpendicular to each other and bisect the angles. Hence, angle OBC = 45° = angle BCO and angle BOC = 90°. The perpendicular OE bisects BC and angle BOC. Thus, BE = $\frac{1}{2}$BC and angle BOE = 45°. The sides and the hypotenuse of a 45° right triangle are 1, 1, and $\sqrt{2}$. Since OB = 4 inches and BE:OB = $1:\sqrt{2}$; then $\frac{BE}{4} = \frac{1}{\sqrt{2}}$ and BE = $\frac{4}{\sqrt{2}} = \frac{4}{2\sqrt{2}} = 2\sqrt{2}$. Therefore, the side of the square is $4\sqrt{2}$ inches.

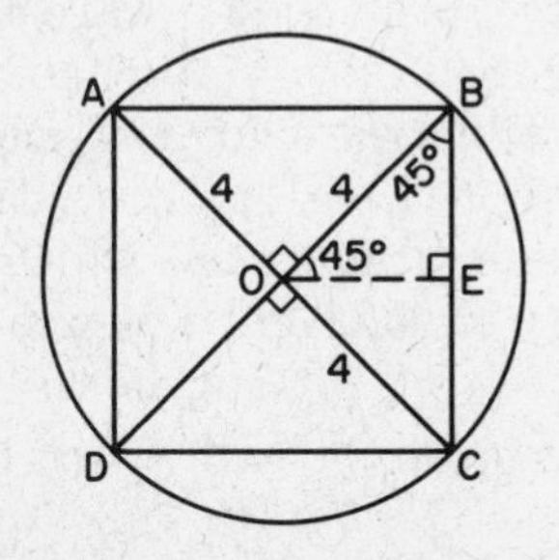

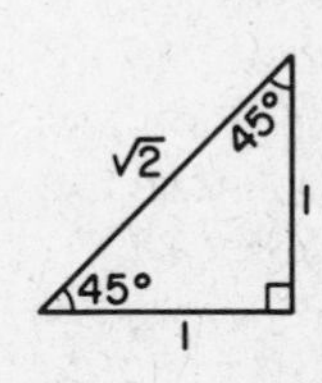

66. a big cheese

55 pounds. Assume a circle containing a diameter of the spherical cheese with radius is OL $= 1$; then the area of the circle is $\pi r^2 = \pi$. Let LP represent the straight cut made by the merchant cutting arc LP $= \frac{1}{4}$ of the circumference. Then PLRS is an inscribed square; OL $=$ OP $=$ OR $= 1$. In right triangle PLR, $(RP)^2 = (RL)^2 + (LP)^2 = 2\,(LP)^2$. But the diameter RP $= 2$, hence $2^2 = 2\,(LP)^2$, or $2 = (LP)^2$, and LP $= \sqrt{2}$. It follows that the area of the shaded segment $= \frac{1}{4}$(area of circle − area of square) $= \frac{1}{4}(\pi - 2)$. Thus, if W equals the weight of cheese, then $\frac{1}{4}(\pi - 2) : \pi = 5 : W$ and $W = 5\pi \div \frac{1}{4}(\pi - 2)$. Assuming $\pi = \frac{22}{7}$ (approx), then W $=$ 55 pounds.

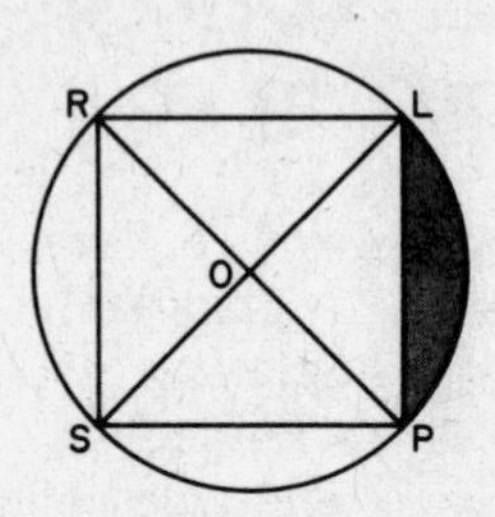

67. said the spider to the fly

Path *c*; 40 feet. Consider the room as if it were a cardboard box cut open and spread on top of a table. The figure shows the three possible ways for the spider to crawl to the fly by a straight path: path *a*, straight across the floor and up the opposite wall, which is 42 feet; path *b*, the hypotenuse of the right triangle SMF, where SM $=$ 37 feet and FM $=$ 17 feet, hence SF $= \sqrt{37^2 + 17^2} = 40.71$ feet (approx); path *c*, the hypotenuse of the right triangle SMF, where SM equals 32 feet and FM = 24 feet, so that the hypotenuse SF = $\sqrt{32^2 + 24^2} = 40$ feet.

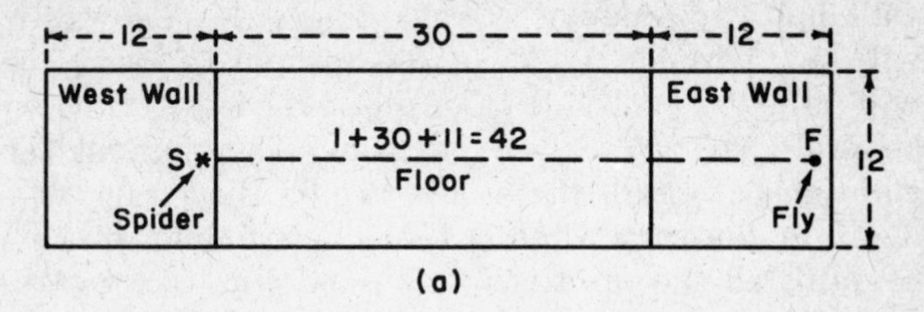

(a)

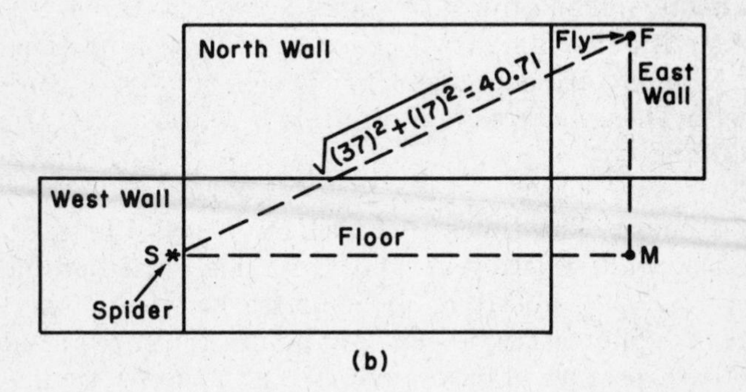

(b)

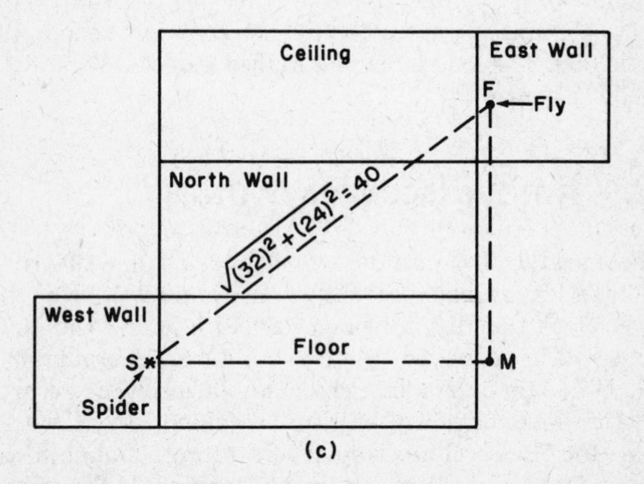

(c)

68. a walking race

a) 3 hours. To be on the same line again, each boy must walk halfway around the track. Therefore, Tony's time will be $(\frac{1}{2} \times \frac{1}{5}) \div 3\frac{1}{2} = \frac{1}{35}$ of an hour; Chris's time will be $(\frac{1}{2} \times \frac{1}{4}) \div 4 = \frac{1}{32}$ of an hour; Paul's time will be $(\frac{1}{2} \times \frac{3}{8}) \div 5 = \frac{3}{80}$ of an hour. The shortest time elapsing before all three boys are on the straight line again at the same time is the smallest number which contains all the fractions $\frac{1}{35}$, $\frac{1}{32}$, and $\frac{3}{80}$ as factors; that is, their least common multiple. To find the least common multiple of $\frac{1}{35}$, $\frac{1}{32}$, and $\frac{3}{80}$ express these fractions with the same least common denominator. Thus, $\frac{1}{35}$, $\frac{1}{32}$, and $\frac{3}{80}$ are equivalent to $\frac{32}{1120}$, $\frac{35}{1120}$, and $\frac{42}{1120}$, respectively. The least common multiple of 32, 35, and 42 is 3360 and $\frac{3360}{1120} = 3$.

b) 6 hours. The times required for Tony, Chris, and Paul to walk entirely around are twice those obtained in *a*, or $\frac{2}{35}$, $\frac{1}{16}$, and $\frac{3}{40}$ of an hour, respectively. As before, the shortest time elapsing before the three are at their respective starting points is the least common multiple of $\frac{2}{35}$, $\frac{1}{16}$, and $\frac{3}{40}$. These fractions are equivalent to $\frac{32}{560}$, $\frac{35}{560}$, and $\frac{42}{560}$, since their least common denominator is 560. The least common multiple of 32, 35, and 42 is 3360; hence, $\frac{3360}{520} = 6$.

69. "time discovers truth"

2 hours and $46\frac{2}{13}$ minutes. In 2 hours from the time the man leaves his office, the minute hand will be at the same place that it was when he had left and the hour hand will have moved $\frac{2}{12}$, or $\frac{1}{6}$, the distance around the dial. Since the hands had changed places on his return, the time over 2 hours will be the time that it would take for the combined movement of both hands to go $\frac{10}{12}$, or $\frac{5}{6}$, of the distance around the dial. Since the

minute hand moves 12 times as fast as the hour hand, the ratio of the distance traveled by the minute hand to that traveled by the hour hand is 12:1; therefore, the minute hand moves over $\frac{12}{13}$ of the distance, or $\frac{12}{13} \times \frac{5}{6} \times 60 = 46\frac{2}{13}$. Therefore, he was away from the office 2 hours and $46\frac{2}{13}$ minutes.

70. "the stuff that life is made of"

a) $55\frac{5}{13}$ minutes. The time spent for lunch is equal to the distance around the dial traveled by the minute hand plus the distance traveled by the hour hand which is equal to 1 complete revolution. Since the minute hand travels 12 times faster than the hour hand, the ratio of the distance traveled by the minute hand to that traveled by the hour hand is 12:1. Since the distance traveled around the dial by the minute hand is $\frac{12}{13}$ of a revolution and the distance traveled by the hour hand is $\frac{1}{13}$ of a revolution, $\frac{12}{13} \times 60 = 55\frac{5}{13}$ minutes.

b) $\frac{60}{143}$ minute past 1 o'clock. Since the minute hand travels 12 times faster than the hour hand, then the distance between them was $\frac{11}{12}$ of the distance of the minute hand from 12:00, the zero point. Hence, the time the man left for lunch is $\frac{11}{12}$ of the distance which is $\frac{1}{13}$ of a revolution. Therefore, the distance is $\frac{12}{11}$ of $\frac{1}{13}$ of a revolution, which is $\frac{12}{11} \times \frac{1}{13} \times 60 = 5\frac{5}{143}$ minutes past 12:00. It then follows that the time when he came back is $5\frac{5}{143} + 55\frac{5}{13} = 60\frac{60}{143}$ minutes after 12:00; that is, $\frac{60}{143}$ of a minute past 1 o'clock.

71. traffic jam

The car. Let $4d$ = the length of the detour in feet. Then, the car has gone $3d$ feet along the detour route, while the truck has gone only a distance of d feet. If it takes the car t minutes to travel $3d$ feet, then it takes the truck $3t$ minutes to travel d feet.

Suppose the car backs up first. Since the car travels backward at one third its forward speed, it will require $3t$ minutes to back up out of the detour. In the meantime, the truck travels forward a distance of $3d$ feet and, since the truck travels at $3t$ minutes for d feet, it will take the truck $3t \times 3 = 9t$ minutes to reach the end of the detour where the car is waiting. If the car starts immediately to cross the detour, since the car travels $3d$ feet in t minutes, it will travel $4d$ feet in $1\frac{1}{3}t$ minutes; hence, this operation would take $9t + 1\frac{1}{3}t = 10\frac{1}{3}t$ minutes.

If the truck backs up first, since the truck travels at one third its forward speed, it will take the truck $3 \times 3t = 9t$ minutes to back up to the beginning of the detour. The car will arrive at the end of the detour at the same time. If the truck starts immediately to cross the detour, since the truck travels d feet in $3t$ minutes, it will travel $4d$ feet in $12t$ minutes; hence, this operation would take $9t + 12t = 21t$ minutes. Therefore, it would be better for the car to back out of the detour first.

72. homework

6 days. Since each day the students will do twice the number of problems already done, when one third of the problems are done at the end of 5 days, the students will do two thirds of the problems the next day; hence, they will complete the assignment in one more day. Therefore, it will take them 6 days.

chapter 4
teasers for the wizard

1. making a number divisible by three

Write a number of as many digits as you please and I shall name a single digit that may be placed before, after, or anywhere within the number making it exactly divisible by 3. Suppose you write 765932. Then I shall put a 1 before, after, or anywhere within the number so that the new number is exactly divisible by 3. Thus, 1765932, 7651932, 7659321, etc. are all exactly divisible by 3.

2. making a number divisible by nine

Write a number of as many digits as you please and I shall name a single digit that may be placed before, after, or anywhere within the number making it exactly divisible by 9. Suppose you write 765932, then I shall put a 4 before, after, or anywhere within the number so that the new number is exactly divisible by 9. Thus, 4765932, 7645932, 7659324, etc. are all exactly divisible by 9.

3. making a number divisible by eleven

Write a number consisting of any number of digits and I shall name an integer that may be added to (or subtracted from) this number that makes it exactly divisible by 11. Suppose you write 765932, then I shall add 9 to this number; thus, $765932 + 9 = 765941$ which is exactly divisible by 11. If you write 639527, then I shall subtract 9 from this number; thus, $639527 - 9 = 639518$ which is exactly divisible by 11.

4. given one digit of a two-digit remainder, reveal the other digit

Choose a number consisting of two digits such that the difference between the two digits is greater than 1 and write this number on a piece of paper without letting anyone see what you have written. Form a new number by reversing the order of the digits and find the difference between the two numbers. Now, if you will tell me one of the digits of the difference, I shall tell you the other digit. For example, write 84 without letting anyone see what you have written. Form the new number 48 by reversing the digits and find the difference, $84 - 48 = 36$. If you will tell me either one of these two digits 6 or 3, I shall name the other digit.

5. given the unit's (or hundred's) digit of a three-digit remainder, reveal the other two digits

Choose any number consisting of three digits such that the hundred's and unit's digits differ by more than 1 and write this number on a piece of paper without letting me see what you have written. Reverse the digits to form a new number and find the difference between these two numbers. Tell me the unit's (or the hundred's) digit in the difference and I shall tell you the other two digits.

Suppose you write 623. Reverse the digits to form a new number 326 and find the difference $623 - 326 = 297$. If you tell me the unit's digit is 7, I shall immediately tell you the ten's digit is 9 and the hundred's digit is 2.

6. revealing the crossed-out digit #1

Write any number without letting me see it and form a new number by reversing the digits. Find the difference between these two numbers and cross out any digit in this difference. Tell me the sum of the remaining digits in the difference and I shall disclose the crossed-out digit.

Suppose you write 5493, form the new number 3945 by reversing the digits, and find the difference $5493 - 3945 = 1548$. Now suppose you cross out the 5. The sum of the remaining digits in the difference is $1 + 4 + 8 = 13$. Tell me this sum 13 and I shall immediately disclose that the crossed-out digit is 5.

7. revealing the crossed-out digit #2

Write any number without letting anyone see it, subtract the sum of its digits, and cross out any digit in the difference. Tell me the sum of the remaining digits in the difference and I shall disclose the crossed-out digit.

Suppose you write 45376. The sum of its digits is $4 + 5 + 3 + 7 + 6 = 25$. Find the difference $45376 - 25 = 45351$. Now suppose you cross out the 3. The sum of the remaining digits is $4 + 5 + 5 + 1 = 15$. Tell me this sum 15 and I shall immediately disclose that the crossed-out digit is 3.

8. lightning calculator #1

Write *two* numbers of four digits each on a piece of paper and I shall quickly write two other numbers of four digits each; then I shall immediately write the sum of the four numbers on another piece of paper. I can also predict the answer and write the sum before you write the two numbers, without letting anyone see it.

You write	5436
	7682
I write	2317
	4563
The sum	19998

I shall now reveal that the number I have written on the other piece of paper is the sum 19998.

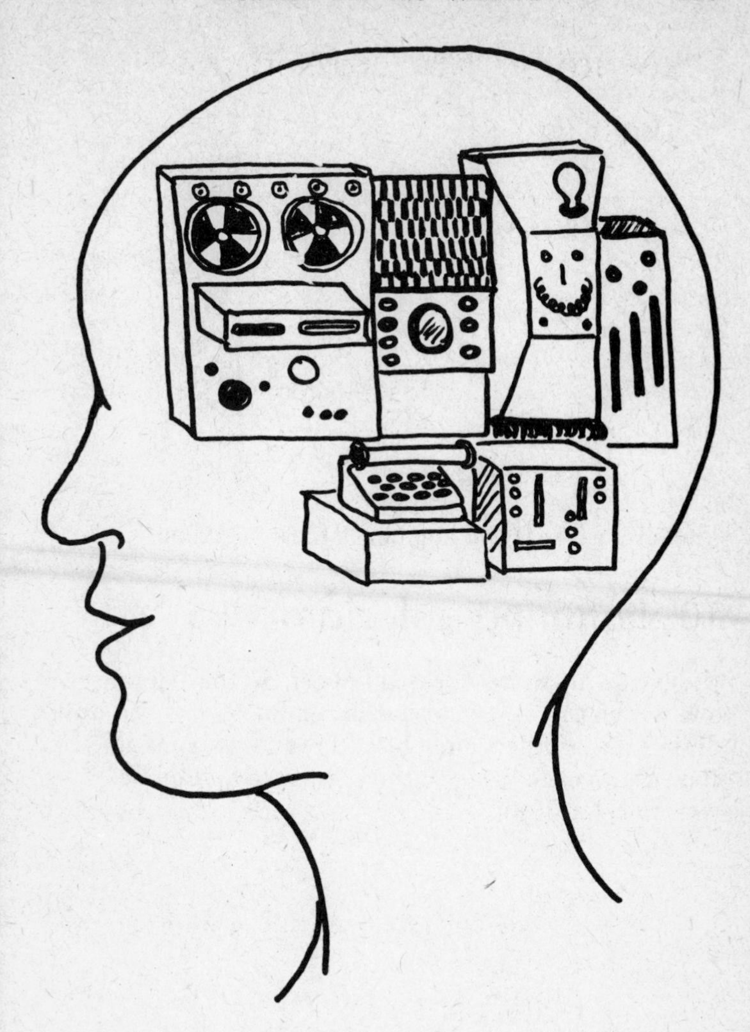

9. lightning calculator #2

I ask you to write four numbers of four digits each, one under the other. I shall then write four numbers of four digits each under the four numbers you have written. Now I can immediately write the sum of all eight numbers,

You write	4327
	5461
	8032
	7253
I write	5672
	4538
	1967
	2746
The sum	39996

I now reveal that the number is the sum 39996.

10. lightning calculator #3

I ask you to write three numbers of four digits each one under the other so that one number has the unit's digit greater then one. I shall write two numbers of four digits each beneath them, then I shall immediately write the total sum.

You write	7123
	2635
	3402
I write	2876
	6597
The sum	22633

11. revealing the remainder

Write any number without letting me see it. Divide this number by 9 and tell me the remainder. Multiply the original number by a digit which I name and divide the resulting product by 9. Then I shall tell you the remainder.

Suppose you write 134 and divide by 9 obtaining the remainder 8. Then multiply 134 by the digit 7, which I name, obtaining the product 938 and divide this product by 9. Without seeing your work, I shall reveal that the remainder is 2.

12. revealing the difference between digits

Write a number containing two digits without letting me see it. Form a new number by reversing the order of the digits and find the difference between the two numbers. Tell me the difference between the two numbers and I shall tell you the difference between the digits of the number you wrote.

Suppose you write 91, obtain 19 by reversing the order of the digits, and tell me the difference $91 - 19 = 72$. I shall immediately reveal that the difference between the two digits of the number that you wrote is 8.

13. telepathy #1

Without letting me see it, write a number of four digits such that the digits decrease by one from left to right. Reverse the order of the digits, obtaining a new number and subtract this number from the original one. Then I, the wizard, can read your mind and, without seeing your work, I can tell you what the difference is.

Suppose you write 7654, reverse the digits to form a new number 4567, and find the difference $7654 - 4567 = 3087$. Without seeing any of your work, I can read your mind and reveal that the difference is 3087.

14. telepathy #2

Without letting anyone see it, write any number of three digits such that the digits decrease by one from left to right. Reverse the order of the digits and obtain a new number, then subtract this number from the original one. Now I, the wizard, can read your mind and, without seeing your work, I can tell you what the difference is.

Suppose you write 765, reverse the digits to form a new number 567, and find the difference $765 - 567 = 198$. Without seeing any of your work I can read your mind and disclose that the difference is 198.

15. revealing an erased digit

Without letting anyone see it, write any integer of three or more digits. Divide by 9 and tell me the remainder. Erase one digit of the original number to form a new number. Divide this new number by 9 and tell me the remainder. I will immediately tell you the digit you erased.

Suppose you write 753, divide 753 by 9, and tell me the remainder 6; then erase the 3 obtaining 75. Now divide 75 by 9 and tell me the remainder, 3; I shall immediately reveal that the digit you erased is 3.

16. telepathy #3

Without letting me see it, write any number of three digits such that the difference between the hundred's and the unit's digits exceeds 1. Form a new number by reversing the order of the digits and subtract it from the original number, finding their difference. Form another number by reversing the order of the digits in the difference; then add this number to the difference. Without seeing your work I, the wizard, can read your mind and tell you the result.

Suppose you write 642, form the new number 246 by reversing the order of the digits, and find the difference $642 - 246 = 396$. Then form new number 693 by reversing the order of the digits in the difference and find the sum $396 + 693 = 1089$. Without seeing any of your work, I can read your mind and tell you that the result is 1089.

17. telepathy #4

Without letting me see it, write any number of four digits such that the digits decrease from left to right, form a new number by reversing the order of the digits, and subtract it from the original number, finding the difference; then form another new number by reversing the order of the digits in the difference and add this number to the previous difference. Without seeing your work I can read your mind and tell you the result.

Suppose you write 8643, form 3468 by reversing the order of the digits, and find the difference $8643 - 3468 = 5175$. Now form another new number 5715 by reversing the order of the digits in the difference and find the sum $5175 + 5715 = 10890$. Without seeing any of your work, I can read your mind and tell you the result is 10890.

18. revealing a number #1

Write any integer, or whole number, on a piece of paper without letting me see what you have written. Square this integer; then square the next larger integer. Tell me the difference between these squares and I shall tell you the number that you originally selected.

Suppose you write 11, square 11 to obtain 121, and square 12 to get 144. Tell me the difference $144 - 121 = 23$ and I shall immediately tell you that the number you selected is 11.

19. revealing a number #2

Write any whole number of two digits without letting anyone see it. Multiply the ten's digit by 2, add 1, multiply by 5 and add the unit's digit. Then tell me the result of these operations and I shall tell you the number you had originally selected.

Suppose you choose 98. Multiply the ten's digit 9 by 2 and add 1; thus, $(9 \times 2) + 1 = 19$, multiply by 5, and add the unit's digit 8; thus, $(19 \times 5) + 8 = 103$. Tell me this result 103 and I shall immediately reveal that the number you selected is 98.

20. revealing a number #3

Without letting me see it, write a number consisting of three digits. Multiply the hundred's digit by 2 and add 1; now multiply this result by 5, add the ten's digit, multiply by 10, and add the unit's digit. Then tell me the result and I shall immediately reveal the number you selected.

Suppose you choose 435. Multiply the hundred's digit by 2, and add 1; thus, $(4 \times 2) + 1 = 9$. Then multiply 9 by 5, add the ten's digit 3, and multiply by 10; thus, $[(9 \times 5) + 3] \times 10 = 480$. Finally, add the unit's digit 5. Now tell me the result $480 + 5 = 485$ and I shall immediately reveal that the number you selected is 435.

21. revealing a number #4

Write any number of two digits without telling me what you have written and square this number. Add 6 to the original number selected and square the result. Now find the difference between the squares. Then tell me the result and I shall reveal the original number selected.

Suppose you write 53, square 53 to obtain 2809, add 6 to 53 to get 59, and square 59 to obtain 3481. Tell me the difference between the two squares $3481 - 2809 = 672$ and I shall reveal that the original number selected is 53.

22. revealing the crossed-out digit #3

Write two or more numbers containing the same number of digits one under the other and I shall write as many numbers under them. I shall then retire, asking you to cross out any digit except 0 in any of the numbers and find the sum of all the numbers, both yours and mine, not adding the crossed-out digit. Now tell me the sum of the digits in the sum that you obtain and I shall reveal the digit that you crossed out.

Suppose you write $\begin{matrix}738\\456\end{matrix}$; then I write $\begin{matrix}261\\543\end{matrix}$ and you cross out the 5 in 456 and add. Thus,

$$\begin{array}{r} 738 \\ 4\not{5}6 \\ 261 \\ 543 \\ \hline 1948 \end{array}$$

The sum of the digits in this sum is $1 + 9 + 4 + 8 = 22$. Tell me this sum 22 and I shall immediately reveal that the crossed-out digit is 5.

23. finding a person's age

If the person is *older* than you:

1) Write down a number containing as many nines as there are digits in your age.
2) Subtract your age.
3) Ask the other person to add the result obtained in 2) to his age without letting you see it.
4) Tell the other person to cancel the left-hand digit and add it to the remaining digits.
5) Ask the other person to tell you the result in 4), and you can immediately reveal his age.

Assume that you are 19; then write down 99 and subtract 19, which leaves 80. Suppose the other person is 21, then he finds the sum $80 + 21 = 101$. Cancel the left-hand digit 1 and add it to 01; thus, $01 + 1 = 2$. Then let him tell you that his result is 2 and you will immediately reveal that his age is 21.

If the other person is *younger* than you, add your age in step 2) and ask him to subtract his age in step 3). Thus, if you are 23, write $99 + 23 = 112$. Suppose he is 19, then $112 - 19 = 103$. Cancel the left-hand digit 1 and add it to 03; thus, $03 + 1 = 4$. Let him tell you that his result is 4 and you will immediately reveal his age is 19.

24. revealing the crossed-out digit #4

Write two or more numbers containing the same number of digits and I shall write a number consisting of as many digits. Then I shall retire, asking you to cross out any digit except 0 in any one of the numbers and find the sum of the numbers (not adding the crossed-out digit). Tell me the sum of the digits in the sum you obtained and I shall reveal the digit which you have crossed out.

You write	4136
	2~~7~~54
	6321
I write	6787
The sum is	19298

Now suppose you cross out the 7 in 2754 and add (not adding the crossed-out digit) to obtain the sum 19298, as shown above. The sum of the digits in the sum is $1 + 9 + 2 + 9 + 8 = 29$. Tell me this sum 29 and I shall immediately reveal that the crossed-out digit is 7.

25. revealing the crossed-out digit #5

Write as many numbers or columns of digits as you like. I shall then write a column consisting of some digits and place this column above, below, on the left, or on the right of the numbers you wrote, wherever you indicate. I shall now retire and ask you to cross out any digit except 0 in any one of these numbers, find the sum of these numbers (not adding the crossed-out digit), and tell me the sum of the digits in this sum. Then I shall immediately reveal the crossed-out digit.

Suppose you write	328	I write	5
	462		6
	5~~7~~2		4
	104		4
	615		6
The sum is	2013		5

Suppose you cross out the 7 in 5724 and add (not adding the crossed-out digit) to obtain the sum 20135, as shown above. The sum of the digits in this sum is $2 + 0 + 1 + 3 + 5 = 11$. Tell me this sum 11, and I shall immediately reveal that the crossed out digit is 7.

In fact, I shall write only a few digits instead of a whole column of digits and place these digits above, below, on the left, or on the right of your numbers, wherever you like.

Suppose this time I write the digits on the left.

		You write	328
			462
I write	6		572
	4		104
	6		615
The sum is	18		011

As before, you cross out the 7 in 6572 and add (not adding the crossed-out digit) to obtain the sum 18011, as shown above. The sum of the digits of this sum is $1 + 8 + 0 + 1 + 1 = 11$. Tell me this sum 11 and I shall immediately reveal that the crossed-out digit is 7.

26. a magic square

The square array of numbers shown below consists of six rows, six columns, and two main diagonals, forming 14 numbers of six digits each.

2	3	4	8	6	4
4	7	2	3	6	5
8	2	4	1	4	8
6	5	8	4	1	3
5	6	2	3	7	4
2	4	7	8	3	3

You may choose any one of the following options and I shall restore the digit, or digits, erased.

1) Erase any digit.
2) Erase any two digits.
3) Erase any three digits.
4) Erase any digit in each column.
5) Erase any digit in each row.
6) Erase any digit in each diagonal.
7) Erase all digits in any column.
8) Erase all digits in any row.
9) Erase all digits in either diagonal.
10) Erase all digits in any column and diagonal.
11) Erase all digits in any row and diagonal.
12) Erase all digits in both diagonals.
13) Rearrange the digits of any row, column, or diagonal, omitting one digit.

27. dad, the wizard

A boy had a box containing eight blocks numbered 1, 2, 3, 4, 5, 6, 7, and 8, respectively. He was playing in his play room with the blocks when he discovered that one block was missing, so he asked his Dad, "One of my blocks is missing; all the blocks are in the play room except blocks 2 and 7, which are in my bed room. Can you tell me which one is missing without leaving your den to look at the blocks?" "Sure," said his father. "Tell Mon to add all the numbers on the blocks in the play room, then tell me the sum of the digits in this sum and I shall give you the number of the missing block." How did the boy's father find the number of the missing block?

28. a guessing game

Have a number of persons select any one of the numbers 1, 2, 3, 4, 5, 6, 7, 8, or 9 and write his choice on a piece of paper, so that no one can see what it is. Then each person takes the eight remaining digits, finds their sum, and then adds the digits in this and succeeding sums repeatedly until he obtains a single digit. Each person then writes the resulting digit on a piece of paper bearing his name. I shall then read the name of the person and reveal the number he selected.

Suppose William Thomas selects 6 and writes it on a piece of paper. He then adds the remaining digits $1 + 2 + 3 + 4 + 5 + 7 + 8 + 9 = 39$ and then writes the sum $3 + 9 = 12$, $1 + 2 = 3$; that is, he writes William Thomas 3 on another piece of paper. I shall then read his name and reveal the number that he selected is 6.

29. revealing the crossed-out digit #6

Without letting me see it, write a number with four digits, find the sum of these digits, form a new number by crossing out any one digit in the original number, and take the difference between the new number and the sum of the digits in the original number. Then tell me the difference and I shall reveal the crossed-out digit.

Suppose you write 8136. The sum of its digits is $8 + 1 + 3 + 6 = 18$. Now, you cross out the 3 in 8136 and obtain 816. Find the difference $816 - 18 = 798$. Tell me this difference 798 and I shall reveal that the crossed-out digit is 3.

30. given one digit of a four-digit number, reveal the other three digits

I turn my back and ask you to write any digit from 1 to 9, inclusive. Multiply this digit by 10, add the original digit, and multiply the result by 9 and then by 11. Tell me the last digit in the result and, with my back still turned, I shall reveal the other three digits.

Suppose you write 7, then multiply by 10, and add the original digit; thus $(7 \times 10) + 7 = 77$. Then multiply by 9 and by 11 to obtain $(77 \times 9) \times 11 = 693 \times 11 = 7623$. Tell me the last digit 3 and I shall reveal that the other three digits in the result are 2, 6, and 7, in this order from right to left.

31. guessing the total thrown with two dice

While I have my back turned, you throw two dice on a table, add the numbers that turn up, and write their sum on a piece of paper. Pick up any one die without disturbing the other and add the number on the opposite face to the sum previously obtained. Throw this one die again and add the number that turns up to the last sum, obtaining a final sum. I then turn around, pick up the dice, and reveal the final sum obtained.

Assume that you throw 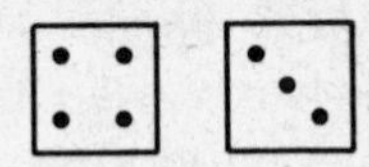

You then write $4 + 3 = 7$ on a piece of paper. Now suppose you pick up the die that turned up 3. the number on the opposite face is 4, so you add $7 + 4 = 11$. Throw this one die again and suppose a 6 turns up; add this 6 to 11 to get 17. I then turn around, pick up the dice, and reveal that the final sum is 17.

32. guessing the throw with two dice #1

While my back is turned throw two dice on a table and keep in mind the two numbers that turn up. Select any one of these numbers, multiply it by 5, add 7 to the product, multiply this sum by 2, and, finally, add to this result the other number that turned up. Tell me this final sum and I shall tell you the two numbers originally thrown.

Suppose you throw 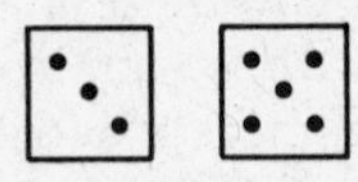

Now suppose you select 3, multiply $3 \times 5 = 15$, add 7 to the product $15 + 7 = 22$, and multiply this sum by 2; thus, $22 \times 2 = 44$. Finally, add to this product the other number 5 that turned up to get $44 + 5 = 49$. Tell me this final sum 49 and I shall reveal that you originally threw a 3 and a 5.

33. guessing the throw with two dice #2

I turn my back and ask you to throw two dice on a table and write the numbers that turn up in order from left to right. Then write to the right of these digits and in the same order the number on the opposite faces of the dice, obtaining a four-digit number, and divide this four-digit number by 11. Tell me the result and I shall reveal the throw and the order of the numbers that turned up.

Suppose you throw 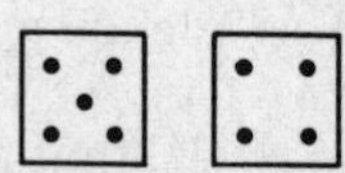

You then write 54. The corresponding numbers on the opposite faces are 23. Thus, you write 5423 and divide 5423 by 11 to obtain 493. Tell me this result and I shall reveal that the original throw was 9 and that the order of the throw was 5, 4.

34. guessing the total thrown with three dice #1

While I have my back turned, throw three dice on a table. Add the numbers that turn up and write the sum on a piece of paper. Pick up any one die without disturbing the others and add the number on the opposite face to the sum previously obtained. Throw this die once again and add the number that turns up to the new sum. I shall then turn around, pick up the dice, and tell you the final sum.

Suppose you throw 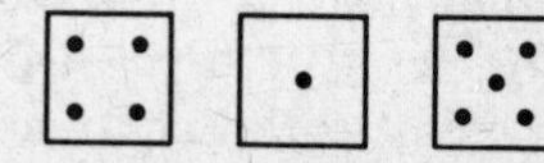

You then write $4 + 1 + 5 = 10$ on a piece of paper and suppose you pick up the middle die. The number on the bottom of this die is 6. Add it to 10, obtaining $10 + 6 = 16$. Throw this die again and suppose you get 4 this time. Add this 4 to 16 to get $16 + 4 = 20$. I turn around, pick up the dice, and reveal that the final sum is 20.

35. guessing the total thrown with three dice #2

While I have my back turned, throw three dice on a table, add the numbers that turn up, and write this

sum on a piece of paper. Pick up any two dice without disturbing the other die and add the numbers on the opposite face of these dice to the sum previously obtained. Throw these two dice again and add the numbers that turn up to the new sum. I shall then turn around, pick up the dice, and tell you the final sum.

Suppose you throw 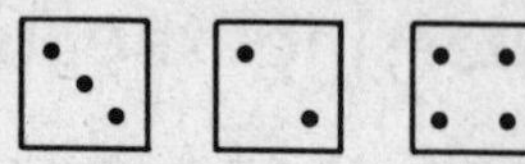

You then write $3 + 2 + 4 = 9$ on a piece of paper and pick up the two dice at the right. The number on the opposite face from the 2 is 5 and the number opposite 4 is 3; add 5 and 3 to 9, obtaining $9 + 5 + 3 = 17$. Throw these two dice again and suppose you get 6 and 1. Add 6 and 1 to 17 to get $17 + 6 + 1 = 24$. I then turn around, pick up the dice, and reveal that the final sum was 24.

36. guessing the throw with three dice #1

I turn my back, ask you to throw three dice on a table, and write the numbers that turn up on a piece of paper. Select any one of the three numbers that turn up. Multiply this number by 2, add 1 to the product, multiply this sum by 5, add any one of the other two numbers that turn up, multiply this sum by 10, add 7, and finally, add the third number that turned up. Tell me the result and I shall tell you the original throw.

Suppose you throw 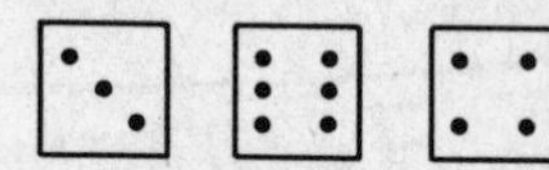

Now suppose you select 4; multiply 4 by 2 and add 1; thus, $(4 \times 2) + 1 = 9$. Multiply this sum 9 by 5; suppose you now select 2 and add to obtain $(9 \times 5) + 2 =$

47. Multiply 47 by 10 and add 7; thus, $47 \times 10 + 7 = 477$. Finally, add the third number 6 to obtain $477 + 6 = 483$. Tell me this final result and I shall reveal that the throw was 2, 4, and 6.

37. guessing the throw with three dice #2

I turn my back, ask you to throw three dice on a table, and write the numbers that turn up in order from left to right. Then to the right of these digits and in the same order write the numbers on the opposite faces of the dice, obtaining a six-digit number. Divide this six-digit number by 37, and then divide the quotient obtained by 3. Tell me the result and I shall reveal the original throw.

Suppose you throw 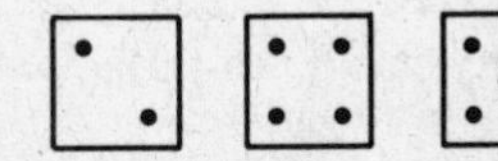

You then write 245; the corresponding numbers on the opposite faces are 532. Thus, write 245532, divide 245532 by 37 to obtain 6636, and divide this quotient 6636 by 3 to get 2212. Tell me this result 2212 and I shall reveal that the original throw was 2, 4, and 5.

38. guessing the total in a column of three dice

I turn my back and ask you to place three dice, one on top of the other, as follows: 1) Place the first die on a table after writing on a piece of paper the number on the face of this die that touches the table, 2) place a second die on top of the first after adding the numbers on the faces of the first and second die that are touching, 3) add this sum to the number on the face of the first die that touches the table, obtaining a new sum, 4) place a third die on top of the second after adding the numbers on the faces of the second and third dice that are touching, and 5) add this sum to the new sum obtained in step 3 to get a final sum. I shall then turn around, cover the column of dice with my hand, and reveal the final sum you obtained.

Assume that 1) the number on the face of the first die touching the table is 4 and 2) the sum of the numbers on the faces of the first and second dice that are touching is $2 + 3 = 5$. 3) Add this sum 5 to the number 4 obtained in step 1, arriving at the new sum $5 + 4 = 9$. 4) The sum of the numbers on the faces of the second and third dice that are touching is $6 + 5 = 11$. 5) Add 11 to the new sum obtained in step 3, getting the final sum $11 + 9 = 20$. I shall then turn around, cover the dice with my hand, and disclose that this sum is 20.

39. guessing the hour #1

I ask you to think of some hour and then to touch another hour on the face of my watch. Then beginning with the hour you touched, start counting to yourself from the hour you thought of and tap each successive hour marked on the face of my watch going in a counterclockwise direction and making the number of taps I tell you to make. When you stop you will be at the hour you thought of.

Suppose you think of IV and touch VIII. Then you tap successively VIII, VII, VI, V, etc. while counting to yourself, 4, 5, 6, 7, etc., making 20 taps. When you stop counting, you will be at the hour you thought of.

40. guessing the hour #2

I ask you to think of some hour. This time, however, I tell you that, beginning at some hour I name, you tap each successive hour marked on the face of the watch the number of taps that I tell you to make going in a counterclockwise direction. Start counting to yourself from the hour you thought of and you will stop at that hour.

Suppose you think of III, and I tell you to begin at IX and tap successively IX, VIII, VII, VI, etc. while counting to yourself 4, 5, 6, 7, etc. until you reach 22. You will then be at III, which is the hour you thought of.

41. guessing the card

I give you a deck of 52 cards, ask you to shuffle them, and then place the deck on a table. You select one of the first 10 cards and remember both the card selected and its number from the top of the deck without revealing this information. Then replace the card in its former position in the deck. I shall pick up the deck, reverse the order of the top 10 cards, and transfer 15 cards from the bottom to the top of the deck. This may be done by dealing the cards. I shall then give the deck back to you and ask you to count to a certain number, beginning with the top card and starting the count with the number following the number indicating the original position of the card you selected. You will then have the card you originally selected.

Suppose you choose the 7th card from the top and it is the 8 of diamonds. You replace the card in its original position, which is the 7th from the top. I then pick up the deck, reverse the order of the top 10 cards, and transfer 15 cards from the bottom to the top by dealing them out. Then I return the deck to you and ask you to count to 26 beginning with the top card and starting the count with the number following 7 which is the number indicating the original position of the card you selected. Thus, the top card is counted 8, the next one, 9, etc. When you reach 26 you will have the ◇8.

42. guessing the total sum of a number of cards

I turn my back and tell you to place a card face down on the table. Form a pile of cards by placing as many other cards as necessary from the deck to make the number of the card at the bottom and the number of cards placed upon it total 12. The honor cards are usually assigned the value 10. Another pile is then formed in the same manner and the operation is repeated until the deck is used up. Then I turn around, ask you for the cards left over, and tell you the sum of the numbers on the bottom cards.

Suppose you place a 5 on the table and then put seven additional cards on top of the 5 to form the first pile. Suppose you form the next pile by placing a 2 on the table and ten additional cards on top of this 2. Then a 4, and eight additional cards to form the third pile; then a 3 and nine additional cards for the fourth pile. Finally an ace and eleven additional cards for the fifth pile. I turn around, ask for the 2 cards left over, and tell you that the sum of the numbers on the bottom cards is 15.

43. revealing a card #1

Take any 16 cards from a deck of 52 and arrange them in four rows and four columns; that is, form a square of four columns each containing four cards.

Without telling me, select and remember any one of the cards. Then I ask you to tell me in which row the selected card lies and note and remember the extreme left-hand card in that row. Beginning with the lowest card in the first row, I pick up the cards in each column face up, one at a time, taking the columns in order from left to right and placing each card on top of the one previously taken up. I then deal the cards in rows of four each beginning with the top left-hand corner. Tell me the row in which the card you selected lies and I shall reveal the card you selected.

Suppose you arrange the 16 cards as shown in the under array (below) and select the 8 of diamonds; you then tell me it is in the third row. I note the extreme left-hand card in this row is the 7 of diamonds and pick up the cards starting with the jack of clubs, then the 7 of diamonds, the 8 of hearts, the king of spades, the ace of hearts, etc., placing each card on top of the one previously taken up. Then I deal the cards in rows of four each beginning with the top left-hand corner as shown in the lower array. Tell me the card that you selected is in the third row and I shall immediately take up the 8 of diamonds.

♠K	♢5	♠6	♣2
♡8	♠9	♢3	♠10
♢7	♡4	♢8	♣9
♣J	♡A	♣Q	♢6

♣J	♢7	♡8	♠K
♡A	♡4	♠9	♢5
♣Q	♢8	♢3	♠6
♢6	♣9	♠10	♣2

44. revealing a pair of cards

Take any 20 cards from a deck of 52 cards and arrange them on a table in ten pairs. Have one, or more persons each select a pair without revealing his selection. I then take up the cards in pairs in any order whatever. Next, I deal these cards out on a table in four rows each containg five cards while I say to myself the magic words, "MUTUS DEDIT NOMEN COCIS." Let anyone tell me the row, or rows, in which the pair he selected is located and I shall immediately reveal the pair he selected.

Assume that you have arranged ten pairs as shown in the upper array (below), and suppose you select (without telling me) the pair consisting of the 3 of diamonds and the king of spades. I then take up the pairs in any order whatever and deal them out on the table, while saying to myself the magic words, "MUTUS DEDIT NOMEN COCIS," as shown in the lower array. If you tell me that the pair you selected is in the second and third rows, I shall immediately tell you that the pair is the 3 of diamonds and the king of spades.

♡Q – ♢Q; ♣3 – ♡K; ♢3 – ♠K; ♡7 – ♠6;
♠3 – ♢6 ; ♡2 – ♣4 ; ♢5 – ♠J ; ♠7 – ♢9;
♣9 – ♡8 ; ♡9 – ♣5

♡9	♡Q	♠3	♢Q	♣9
♣3	♢3	♡K	♢9	♢6
♢5	♡7	♣5	♠K	♠J
♡2	♠6	♣4	♠7	♡8

45. revealing a card #2

I take 27 cards from a deck of 52 and hold these cards in my hand facing upward. I deal the cards face up on a table in three piles of nine cards each as follows: The top card of the pack that I hold in my hand is dealt as the bottom card of the first pile, the second card of the pack of 27 cards is dealt as the bottom card of the second pile, the third card is dealt as the bottom card of the third pile, the fourth card is dealt as the second card of the first pile, the fifth as the second card of the second pile, and so on.

As I deal the cards, I ask you to choose a card, without letting me know your choice, and remember in which pile it was located. After I finish dealing the cards, I ask you to tell me in which pile your choice is located. I then take up the three piles placing the pile you indicated between the other two. I deal again, as before, and again ask you to tell me in which pile your choice is located. Then I pick up the three new piles, placing the pile you indicated between the other two. Finally, I deal the cards in the same manner as before and note the middle, or fifth, card in each pile. If you then tell me which pile contains your choice, I shall tell you which card it was.

46. the christians and turks

Josephus, the author of a Jewish History, escaped death by working a problem similar to this one. When the Roman emperor Vespasian captured Jotapat, Josephus with forty other Jews had to hide in a cave. All those in the cave decided to kill themselves rather than fall into the hands of their conquerors. Not wishing to go along with this decision, Josephus persuaded the others that they should die by lot. He then suggested that they should arrange themselves in a circle and that every third person should be killed. He arranged all those in the cave so that all of them were killed except himself and another whom he might easily destroy or persuade to yield to the Romans. In the middle ages, this problem took the following form: Fifteen Christians and fifteen Turks were carried as passengers in a ship when they met a violent storm. It was decided that to save the ship and crew, half of the passengers had to be thrown overboard. To carry this out equitably it was agreed that the passengers be placed in a circle and every ninth man be cast into the sea until one half of them had been thrown overboard. The captain, being a Christian, arranged them so that all Christians were saved. How did he arrange them?

47. revealing a selected digit and the age of a person

Choose any digit between 1 and 10, double this digit, and add 5; then multiply the result by 50, add 1719, and substract the year in which you were born. Tell me the result of your calculations and I shall reveal the digit which you had chosen, as well as your age.

Suppose you choose 7 and you were born in 1928, then the operations performed are $7 \times 2 = 14$, $14 + 5 = 19$, $19 \times 50 = 950$, $950 + 1719 = 2669$, and, since you were born in 1928, then $2669 - 1928 = 741$, which is your final result. Tell me this result and I shall reveal that the digit that you had chosen is 7, and that you are 41 years old.

solutions

1. making a number divisible by three

The sum of the digits of 765932 is $7 + 6 + 5 + 9 + 3 + 2 = 32$. Find the difference between 32 and the nearest integral multiple of 3 which is greater than 32. Thus, the nearest integral multiple of 3 greater than 32 is 33 and the difference, 33 − 32 = 1. The digit 1 can be placed before, after, or anywhere within the number 765932 and the resulting number is divisible by 3. Thus, 1765932, 7651932, 7659321, etc. are divisible by 3.

Explanation. A number is divisible by 3 if the sum of its digits is divisible by 3.* Hence, a digit must be annexed to the given number such that the sum of the digits of the new number is divisible by 3. This digit is the difference between the sum of the digits of the given number and the nearest integral multiple of 3, which is greater than the number.

2. making a number divisible by nine

The sum of the digits of 765932 is 7 + 6 + 5 + 9 + 3 + 2 = 32. Find the difference between 32 and the nearest integral multiple of 9, which is greater than 32. Thus, the nearest integral multiple of 9 greater than 32 is 36. The difference, 36 − 32 = 4, hence a 4 can be placed before, after, or anywhere within the number 765932 and the resulting number is divisible by 9. Thus, 4765932, 7654932, 7659324, etc. are divisible by 9.

Explanation. A number is divisible by 9 if the sum of its digits is divisible by 9. Hence, a digit must be annexed to the given number such that the sum of the digits of the new number is divisible by 9. This digit is the difference between the sum of the digits of the given number and the nearest integral multiple of 9, which is greater than the number.

* For a proof of the divisibility of a number by 3, 9, or 11, see Julio A. Mira, *Arithmetic Clear and Simple* (New York: Barnes & Noble, Inc., 1965), p. 159.

3. making a number divisible by eleven

The given number is 765932. The sum of the digits in the odd places is $2 + 9 + 6 = 17$. The sum of the digits in the even places is $3 + 5 + 7 = 15$. If the sum of the digits in the odd places is greater than the sum of the digits in the even places, as in this case, *add* the *difference* between these results, $17 - 15 = 2$ and the nearest integral multiple of 11, which is greater than 2. That is, $11 - 2 = 9$ is added to the given number; thus, $765932 + 9 = 765941$ which is divisible by 11.

If you write 639527, the sum of the digits in the odd places is $7 + 5 + 3 = 15$ and the sum of the digits in the even places is $2 + 9 + 6 = 17$. The difference is $17 - 15 = 2$. If the sum of the digits in the odd places is less than the sum of the digits in the even places, *subtract* the *difference* between this result $17 - 15 = 2$ and the nearest integral multiple of 11, which is greater than 2; that is, $11 - 2 = 9$ from the given number; thus, $639527 - 9 = 639518$ which is divisible by 11.

Explanation. Any number divided by 11 leaves the same remainder as the difference between the sum of its digits in the odd places and those in the even places divided by 11. It then follows that a number is divisible by 11 if the difference between the sum of its digits in the odd places and those in the even places is either divisible by 11 or equal to 0.

If the sum of the digits in the odd places is *greater* than the sum of the digits in the even places, the difference between these sums is the remainder after dividing by 11. If we add to the given number the difference between the remainder and the nearest integral multiple of 11, which is greater than this difference, then upon dividing the new number by 11, it will have a remainder of 0 and, therefore, is divisible by 11.

If the sum of the digits in the odd places is *less* than the sum of the digits in the even places, the difference between these sums is the amount that the remainder lacks from being 11. Thus, if we subtract from the given number the difference between the amount that the remainder lacks from being 11 and the nearest integral multiple of 11, which is greater than this difference, then upon dividing the new number by 11, it will have a remainder of 0; thus, it is divisible by 11.

4. given one digit of a two-digit remainder, reveal the other digit

If you write 84, then the new number is 48. The difference is $84 - 48 = 36$. Suppose you name the unit's digit 6, then the other digit is the difference between 9 and this digit, which is $9 - 6 = 3$; hence, the number is 36.

Explanation. Any number divided by 9 leaves the same remainder as the sum of its digits. Since the new number is formed by reversing the digits of the chosen number, the two numbers consist of the same digits. Thus, the sum of the digits of each of the two numbers is the same; that is, they leave the same remainder upon division by 9. Now, if n_1 and n_2 denote the two numbers and q_1 and q_2 the respective quotients, then we can let r represent the remainder in each case, since it is the same for both. It follows that,

$$\begin{array}{r} n_1 = 9q_1 + r \\ n_2 = 9q_2 + r \\ \hline n_1 - n_2 = 9(q_1 - q_2) \end{array}$$

That is to say, the difference $n_1 - n_2$ is divisible by 9. Hence, the sum of the digits of the difference is divisible by 9, so that the sum of the digits in the difference must equal 9 or some integral multiple of 9.

5. given the unit's (or hundred's) digit of a three-digit remainder, reveal the other two digits

If you write 623, then the new number is 326. The ten's digit will always be the same for both numbers; hence, the ten's digit in the difference $623 - 326 = 297$ will always be 9. Assume you name the unit's digit 7. The hundred's digit is $9 - 7 = 2$ and the number is 297.

Explanation. The number chosen and the new number formed by reversing the digits must have the same ten's digit. It follows that the difference between these two numbers must have 9 in the ten's place. By problem 4, the difference is divisible by 9; that is, the sum of its digits is 9 or some integral multiple of 9. Since the ten's digit is 9, the sum of the other two digits must equal 9. Therefore, the ten's digit is 9, and the hundred's digit is the difference between 9 and the unit's digit.

6. revealing the crossed-out digit #1

If the number taken is 5493, then the new number is 3945 and the difference is $5493 - 3945 = 1548$. Suppose 5 is the digit crossed out in this difference. The sum of the remaining digits $1 + 4 + 8 = 13$. Then the crossed-out digit is the difference between this sum and the nearest integral multiple of 9 which is greater than 13. Thus, $18 - 13 = 5$, which is the crossed-out digit.

Explanation. The difference between two numbers consisting of the same digits is divisible by 9. (See problem 4.) Hence, the sum of the digits of the difference is divisible by 9. That is, the sum of the digits of the difference must be 9 or some integral multiple of 9. Therefore, when a digit is crossed out from the difference, the sum of the remaining digits differs from 9 or some integral multiple of 9 by the crossed-out digit.

7. revealing the crossed-out digit #2

The number chosen is 45376. The sum of its digits is $4 + 5 + 3 + 7 + 6 = 25$ and the difference is $45376 - 25 = 45351$. If the 3 is crossed out, the sum of the remaining digits is $4 + 5 + 5 + 1 = 15$. Then the crossed-out digit is the difference between 15 and the nearest integral multiple of 9 which is greater than 15. Thus, $18 - 15 = 3$ which is the crossed-out digit.

Explanation. The sum of the digits of a number is equal to the remainder upon division by 9. Hence, if the sum of the digits of a number is subtracted from the number, the difference will have a remainder of zero upon division by 9. That is to say, the difference is exactly divisible by 9, so that the sum of the digits of this difference is equal to 9 or some integral multiple of 9. Therefore, when a digit is crossed out from the difference, the sum of the remaining digits differs from 9 or some integral multiple of 9 by the crossed-out digit.

8. lightning calculator #1

The number I write on a piece of paper before the person writes the two numbers is $19998 = 9999 \times 2$.

Explanation. After you write two numbers, I write the other two numbers such that each corresponding pair of digits adds up to 9. Thus, corresponding to 7682, I write $9 - 7 = 2$, $9 - 6 = 3$, $9 - 8 = 1$ and $9 - 2 = 7$; that is, 2317. Similarly, the number corresponding to 5436 is 4563. Hence, I have two pairs of numbers,

$$\begin{matrix} 7682 \\ 2317 \end{matrix} \text{ and } \begin{matrix} 5436 \\ 4563 \end{matrix}$$

each of which adds up to 9999. Therefore, the result must be $9999 \times 2 = 19998$.

9. lightning calculator #2

The number I write is $39996 = 9999 \times 4$.

Explanation. After you write the four numbers, I write the other four numbers such that each corresponding pair of digits adds up to 9. Thus, the number corresponding to 4327 is obtained as follows: $9 - 4 = 5$, $9 - 3 = 6$, $9 - 2 = 7$, and $9 - 7 = 2$; hence, the number corresponding to 4327 is 5672, etc. I then have four pairs of such numbers,

4327	5461	8032	7253
5672	4538	1967	2746

Each of these adds up to 9999; therefore, the result is $9999 \times 4 = 39996$.

10. lightning calculator #3

Choose any two of the numbers so that the remaining number has the unit's digit greater than 1. Suppose you choose 7123, 3402, and 2635. Then I write two other numbers such that each corresponding pair of digits adds up to 9. Thus, corresponding to 7123, I write $9 - 7$

$= 2$, $9 - 1 = 8$, $9 - 2 = 7$, and $9 - 3 = 6$; that is, 2876. Similarly, corresponding to 3402, I write 6597. The remaining number is 2635. Take 2 away from the unit's digit and prefix 2 to this number to obtain the answer 22633.

Explanation. I have two pairs $\frac{7123}{2876}$ and $\frac{3402}{6597}$, each of which adds up to 9999. Hence, the sum of these four numbers is $9999 \times 2 = 19998$. It is easily seen that, since the unit's digit 8 lacks 2 from being 10, the result of adding any digit greater than 1 to 8 is always 2 less than the digit added. Moreover, because in each case we carry one, all the other digits will be the same as those of the number added, for each time we are adding 10 to each digit. Finally, the ten thousand's digit must be 2. Therefore, since the remaining number is 2635, the answer is 22633.

11. revealing the remainder

The number 134 divided by 9 gives a remainder equal to $1 + 3 + 4 = 8$. (See problem 4.) The product $134 \times 7 = 938$ and if we divide this product 938 by 9, the remainder is $9 + 3 + 8 = 20$, or $2 + 0 = 2$. The product $8 \times 7 = 56$ and the remainder when 56 is divided by 9 is $5 + 6 = 11$ and $1 + 1 = 2$.

Explanation. Any number N divided by 9 is equal to $9q + r$, where q is the quotient and r is the remainder; that is, $N = 9q + r$. Multiplying N by any other number d gives,

$$N \times d = d\ (9q + r) = 9dq + (dr).$$

Thus, the remainder upon dividing the product $N \times d$ by 9 is the product dr, which is the product of the first remainder by the number we chose for the multiplier.

12. revealing the difference between digits

Divide the difference $91 - 19 = 72$ by 9 obtaining 8. This number 8 is the difference between the two digits 9 and 1.

Explanation. Let h represent the hundred's digit and u, the unit's digit.

The number is	$10h + u$
The new number is	$10u + h$
The difference is	$9h - 9u = 9(h - u)$

Thus, if the difference between the two numbers is denoted by d, then

$$9(h - u) = d \text{ and } h - u = \frac{d}{9}$$

That is, the difference between the two digits $h - u$ is equal to the difference between the two numbers d divided by 9.

13. telepathy #1

$7654 - 4567 = 3087$. No matter what number of four digits you write such that the digits decrease by one from left to right, the difference between this number and the number formed by reversing the digits is always 3087.

Explanation. If a represents the thousand's digit, then $a - 1$ is the hundred's digit, $a - 2$ is the ten's digit and $a - 3$ is the unit's digit. The number is

$$1000a + 100(a-1) + 10(a-2) + (a-3)$$

Reversing the digits, the new number is,

$$1000(a-3) + 100(a-2) + 10(a-1) + a$$

The difference is

$$\begin{array}{lllll} 1000a & +100(a-1) & +10(a-2) & +a-3 & \\ 1000(a-3) & +100(a-2) & +10(a-1) & +a & \\ \hline 3000 & +100 & +(-10) & +(-3) & \\ & & & & =3087 \end{array}$$

That is, no matter what number of four digits you write, such that the digits decrease by one from left to right, the difference between this number and the number formed by reversing the digits is always 3087.

14. telepathy #2

$765 - 567 = 198$. No matter what number of three digits you write, where the digits decrease by one from left to right, the difference between this number and the number formed by reversing the digits is always 198.

Explanation. If a represents the hundred's digit, then $a - 1$ is the ten's digit and $a - 2$ the unit's digit. So that the number is

$$100a + 10(a-1) + (a-2)$$

Reversing the digits, the new number is,

$$100(a-2) + 10(a-1) + a$$

The difference is

$$\begin{array}{llll} 100a & +10(a-1) & +(a-2) & \\ 100(a-2) & +10(a-1) & +\ a & \\ \hline 200 & +\ \ 0 & +(-2) & =198 \end{array}$$

That is, no matter what number of three digits you write, such that the digits decrease by one from left to right, the difference between this number and the number formed by reversing the digits is always 198.

15. revealing an erased digit

If the second remainder is less than the first, take the difference between them. Thus, if you divide 753 by 9, the remainder is 6. Then if you erase the 3 and divide 75 by 9, the remainder is 3. The figure you erased is $6-3=3$. If the second remainder is greater than the first, subtract the difference between them from 9. Thus, if you divide 731 by 9, the remainder is 2. Then, if you erase the 7 and divide 31 by 9, the remainder is 4. The difference $4-2=2$ and the figure erased is $9-2=7$. If the remainders are equal, then the digit erased is either 9 or 0.

Explanation. Any number divided by 9 leaves the same remainder as the sum of its digits. (See problem 4.) In obtaining the first remainder you get the sum of the digits of the original number. When you erase a digit and divide the new number by 9, you obtain the sum of the digits of the original number less the digit you erased. It follows that the difference between the two remainders is the digit you erased.

16. telepathy #3

No matter what number of three digits you choose, the result of these operations is always 1089.

Explanation. Let h, t, and u denote the hundred's, ten's, and unit's digits, respectively, where h is greater than u by more than 1.

The original number is $100h + 10t + u$
The new number is $100u + 10t + h$

To subtract h from u, I must borrow one unit of 10 from $10t$, for u is less than h.

The original number is $100h + 10(t-1) + (u+10)$
The new number is $100u + 10t + h.$

To subtract $10t$ from $10(t-1)$, I must borrow one unit of 100 from $100h$, for $10(t-1)$ is less than $10t$.

The original number is
$$100(h-1) + [10(t-1) + 100] + (u+10)$$
The new number is
$$100u + 10t + h$$

The difference is
$$100(h-1-u) + 10(9) + (u+10-h)$$
The new number is
$$100(u+10-h) + 10(9) + (h-1-u)$$

The sum is
$$900 + 180 + 9 = 1089$$

That is, no matter what number of three digits you choose such that the hundred's and the unit's digits differ by more than 1, the result of these operations is always 1089.

17. telepathy #4

No matter what number of four digits you choose such that the digits decrease from left to right, the result of these operations is always 10890.

Explanation. Let a, h, t, and u denote the thousand's, hundred's, ten's, and unit's digits, respectively, where a is greater than h, h is greater than t, and t is greater than u.

The original number is $1000a + 100h + 10t + u$
The new number is $1000u + 100t + 10h + a$

Now, a is at least three more than u, so we must borrow one unit of 10 from t.

The original number is

$$1000a + 100h + 10(t - 1) + (u + 10)$$

The new number is

$$1000u + 100t + 10h \qquad + a$$

But h is greater than t, hence h is greater than $t - 1$, so we must borrow one unit of 100 from h.

The number is

$$1000a + 100(h - 1) + 10(t - 1 + 10) + (u + 10)$$

The new number is

$$1000u + 100t \qquad + 10h \qquad + a$$

Since h is greater than t by at least 1 then $h - 1$ is either greater than or at least equal to t. Hence, we do not have to borrow from a

The difference is

$$\begin{array}{llll} 1000(a - u) & + 100(h - t - 1) & + 10(t - h + 9) & + (u + 10 - a) \\ 1000(u + 10 - a) & + 100(t - h + 9) & + 10(h - t - 1) & + (a - u) \\ \hline 10000 & + 800 & + 80 & + 10 \end{array}$$

That is, no matter what number of four digits you choose such that the digits decrease from left to right, the result of these operations is always 10890.

18. revealing a number #1

The difference between the squares of 11 and 12 is $144 - 121 = 23$. Subtract 1 from this difference and divide by 2. The result $(23 - 1) \div 2 = 11$, which is the number you selected.

Explanation. If n denotes the integer selected, then $n + 1$ is the next larger integer. So that $(n + 1)^2 - n^2$ is the difference d; thus, $(n + 1)^2 - n^2 = d$, or $n^2 + 2n + 1 - n^2 = d$; that is, $2n + 1 = d$, or $n = \frac{d - 1}{2}$. Therefore, the integer selected is always equal to one less than the difference between their squares divided by 2.

19. revealing a number #2

Suppose you select 98, the result of the operations performed yield $(9 \times 2) + 1 = 19$, then $(19 \times 5) + 8 = 103$. To obtain the number selected, subtract 5 from this result; thus, $103 - 5 = 98$.

Explanation. Let t and u denote the ten's and the unit's digits, respectively, then the number selected is $10t + u$. If you multiply the ten's digit t by 2 and add 1, you obtain $2t + 1$. If then you multiply $2t + 1$ by 5 and add the unit's digit u, you will obtain $5(2t + 1) + u = 10t + u + 5$; that is, the result is 5 more than the number selected. Therefore, if 5 is subtracted from the result of these operations, you will obtain the number selected.

20. revealing a number #3

If the number selected is 435, the operations performed yield $(4 \times 2) + 1 = 9$; $(9 \times 5) + 3 = 48$; $(48 \times 10) + 5 = 485$; subtract 50 from this result and the difference $485 - 50 = 435$ is the number selected.

Explanation. Let h, t, and u denote the hundred's, ten's and unit's digits, respectively. Then the number selected is $100h + 10t + u$. If you multiply the hundred's digit h by 2 and add 1, you will obtain $2h + 1$. Then multiply this result by 5, add the ten's digit, and multiply by 10. Thus, $[5(2h + 1) + t]\,10 = 100h + 50 + 10t$. Finally, add the unit's digit and you have $100h + 10t + u + 50$. That is, the result is 50 more than the number selected. Therefore, if you subtract 50 from the result of these operations, you will obtain the number selected.

21. revealing a number #4

If the number selected is 53, the operations performed yield $(53 + 6)^2 - (53)^2 = (59)^2 - (53)^2 = 3481 - 2809 = 672$. Subtract 36 from this difference and divide by 12; thus, $672 - 36 = 636$ and $636 \div 12 = 53$, which is the number selected.

Explanation. Let n denote the number selected. The operations to be performed yield $(n + 6)^2 - n^2 = n^2 + 12n + 36 - n^2 = 12n + 36$. Thus, if d represents the difference between the squares, then $12n + 36 = d$, or $n = \dfrac{d - 36}{12}$. Therefore, the integer selected is always equal to 36 less than the difference between the squares divided by 12.

22. revealing the crossed-out digit #3

Subtract the sum 22 from the nearest integral multiple of 9 which is greater than 22; that is, 27. Thus, $27 - 22 = 5$, which is the crossed-out digit.

Explanation. No matter what numbers you write, I shall write the other numbers such that each corresponding pair of digits adds up to 9. Thus, corresponding to 738. I write $9 - 7 = 2$; $9 - 3 = 6$; $9 - 8 = 1$, or 261. Similarly, the number corresponding to 456 is 543. Hence, I have two pairs of numbers $\begin{matrix}738\\261\end{matrix}$ and $\begin{matrix}456\\543\end{matrix}$ each of which adds up to $9999 = 9 \times 1111$. Thus, each pair is exactly divisible by 9, hence their sum is exactly divisible by 9. Since the sum is exactly divisible by 9, then the sum of the digits of this sum is also exactly divisible by 9. (See problem 2.) Therefore, the digit crossed out is obviously the difference between the sum of the remaining digits and the nearest integral multiple of 9, which is greater than this sum.

23. finding a person's age

Suppose you are 19 and the other person is *older*. Add the result 2 to your age 19; thus, $19 + 2 = 21$, which is his age. If the other person is *younger* and you are 23, subtract his result 4 from your age 23; thus, $23 - 4 = 19$, which is his age.

Explanation. Let a denote your age and b represent the other person's age. If he is *older* than you, the operations yield $99 - a + b - 100 + 1 = -a + b = b - a$; that is, the result $b - a$ is the difference in your ages. Hence, if you add $b - a$ to your age, you will obtain $b - a + a = b$ which is his age.

If the other person is *younger*, the operations yield $99 + a - b - 100 + 1 = a - b$; that is, the result $a - b$ is the difference in your ages. Hence, if you subtract $a - b$ from your age, you will obtain $a - (a - b) = a - a + b = b$, which is his age.

24. revealing the crossed-out digit #4

Subtract the sum of the digits 29 from the nearest integral multiple of 9 which is greater than 29; that is, 36. Thus, $36 - 29 = 7$ is the crossed-out digit.

Explanation. I write the other number such that the sum of the digits in each column written by you is 9 or any integral multiple of 9. Thus, the sum of the unit's digits is $6 + 4 + 1 = 11$, and $11 + 7 = 18$; hence, I write 7 for my unit's digit. Similarly, the sum of the ten's digits written by you is $3 + 5 + 2 = 10$, and $10 + 8 = 18$. Thus, I write 8 for the ten's digit. Then the sum of the hundred's digits written by you is $1 + 7 + 3 = 11$, and $11 + 7 = 18$, so I write 7 for the hundred's digit. Finally, the sum of the thousand's digits written by you is $4 + 2 + 6 = 12$ and $12 + 6 = 18$, so I write 6 for the thousand's digit; that is, the number I write is 6787. It follows that the sum of these numbers contains 9 as a factor; hence, it is divisible by 9. Therefore, the sum of the digits in this sum must be divisible by 9. (See problem 2.) When you cross out a digit, you obtain the sum of the digits of the original numbers less the crossed-out digit. Therefore, the difference between the sum of the digits in the sum you obtain and the nearest integral multiple of 9 which is greater than this sum is the crossed-out digit.

25. revealing the crossed-out digit #5

Subtract the sum of the digits 11 from the nearest integral multiple of 9 that is greater than this sum, which is 18. Thus, $18 - 11 = 7$ is the crossed-out digit.

Explanation. I write the column such that the sum of the digits in each row is 9 or an integral multiple of 9; thus,

$$
\begin{array}{lll}
328 & 3+2+8=13; & 18-13=5 \\
572 & 5+7+2=14; & 18-14=4 \\
462 & 4+6+2=12; & 18-12=6 \\
104 & 1+0+4=\ \ 5; & \ \ 9-\ \ 5=4 \\
615 & 6+1+5=12; & 18-12=6
\end{array}
$$

Now, each row contains 9 as a factor; hence, the sum of the numbers contains 9 or some multiple of 9 as a factor, and the sum of the digits of this sum is divisible by 9. (See problem 2.) Since I want to make the sum of the digits of the sum of the numbers 9 or some integral multiple of 9, I can cancel out any combination of numbers (in the column I write) that add to 9. Thus, I don't have to write the 5 and 4, but simply,

$$
\begin{array}{r}
328 \\
5\not{7}2 \\
4626 \\
1044 \\
\underline{6156} \\
20126
\end{array}
$$

The digits in the sum less the crossed-out digit add up to $2+0+1+2+6=11$. Therefore the difference between this sum and the nearest integral multiple of 9 that is greater than 11 is $18-11=7$, which is the crossed-out digit. It is easily seen that it makes no difference whether I write the column at the left, the right, below, or above, for, in any case, the sum of the digits of the sum will be a multiple of 9.

26. a magic square

The sum of the digits in each column, row, and diagonal is 27. All the options can be reduced to the sum of the digits of a column, row, or diagonal. Hence, the missing digit can be found by adding the remaining digits and subtracting from 27.

Suppose you erase 8 in the fourth row and 2 and 3 in fifth row, we can find 8 by adding all the other digits in the fourth row $6 + 5 + 4 + 1 + 3 = 19$ and subtracting $27 - 19 = 8$. Once 8 is found, we can find 2 by adding all the other digits in the third column $4 + 2 + 4 + 8 + 7 = 25$ and finding the difference $27 - 25 = 2$. Finally, we can find 3 by adding all the other digits in in either the fourth column or the fifth row. Thus, using the fourth column $8 + 3 + 1 + 4 + 8 = 24$ and $27 - 24 = 3$.

27. dad, the wizard

The sum of the numbers $1 + 2 + 3 + 4 + 5 + 6 + 7 + 8 = 36$ which is a multiple of 9. Since the missing blocks are numbered 4 and 5 and $4 + 5 = 9$, then the sum of the remaining numbers is a multiple of 9. The boy's dad subtracted the sum of the digits of the sum which the boy gave him from the nearest integral multiple of 9 greater than the sum. Thus, if the missing block is 6, the sum given by the boy is $1 + 2 + 3 + 7 + 8 = 21$. The difference between 21 or $2 + 1 = 3$ and the nearest integral multiple of 9 greater than 21 or 3 is either $27 - 21 = 6$ or $9 - 3 = 6$, which is the number of the missing block.

28. a guessing game

Subtract the digit written on the paper from 9. Thus, if the paper I read has the digit 3 written on it, then the digit selected is $9 - 3 = 6$.

Explanation. The sum of the numbers $1 + 2 + 3 + 4 + 5 + 6 + 7 + 8 + 9 = (1 + 9) + (2 + 8) + (3 + 7) + (4 + 6) + 5 = 45$ and $4 + 5 = 9$. Hence, this sum is exactly divisible by 9. (See problem 2.) That is to say, the remainder upon dividing this sum by 9 is 0. (See problem 4.) If you select one of the digits in this sum and add the remaining digits you will obtain a new sum; the sum of the digits in this new sum is the remainder obtained upon dividing this sum by 9. It follows that the sum of the digits of this new sum differs from 9 by the digit selected.

29. revealing the crossed-out digit #6

If the number written is 8136 and you cross out the 3, the difference is $816 - 18 = 798$. Find the sum of the digits of this number; thus, $7 + 9 + 8 = 24$ and $2 + 4 = 6$. Subtract this final sum from 9 to obtain $9 - 6 = 3$, the crossed-out digit.

Explanation. Let a, b, c, and d denote the thousand's, hundred's, ten's and unit's digits, respectively, of the number selected. The sum of the digits of this number is $a + b + c + d$. Suppose you cross out the hundred's digit b, then the new number is $100a + 10c + d$. The difference between this new number and the sum of the digits of the original number is $100a + 10c + d - (a + b + c + d) = 100a + 10c + d - a - b - c - d$, or $(100 - 1)\,a + (10 - 1)\,c - b = 99a + 9c - b$. If this number is divided by 9, the remainder is $-b$; that is, the remainder lacks b units of being 9, so that it is $9 - b$. Now, if we subtract this remainder from 9, the result is $9 - (9 - b) = 9 - 9 + b = b$, which is the crossed-out digit.

30. given one digit of a four-digit number, reveal the other three digits

Assuming the number you obtain is 7623 and you tell me the unit's digit is 3, then the ten's digit is $3 - 1 = 2$, the hundred's digit is $9 - 3 = 6$, and the thousand's digit is $9 - 2 = 7$.

Explanation. Let n denote the digit selected. The indicated operations yield $(10n + n)\ 9 \times 11 = 1089n$. This number $1089n$ is exactly divisible by both 9 and 11. The operation $1089 \times n$ yields as its first product $9n$. Now, $9 \times 2 = 18$, $9 \times 3 = 27$, $9 \times 4 = 36$, etc. That is, multiplying 9 by any digit n between 1 and 9, inclusive, will result in carrying one less than the multiplier n which is $n - 1$. The second product is then $8n + (n - 1) = 9n - 1$. Hence, the second product is one less than the first product, therefore the ten's digit is one less than the unit's digit in the resulting number.

Let a, h, t, and u denote the thousand's, hundred's, ten's, and unit's digits, respectively, of the resulting number, then $t = u - 1$, because the resulting number is exactly divisible by 9, and the sum of its digits must equal 9 or some integral multiple of 9. Thus, $a + h + u - 1 + u = a + h + 2u - 1$ equals 9 or some integral multiple of 9. Moreover, since the resulting number is also divisible by 11, the difference between the sum of its hundred's and unit's digits, and the sum of its thousand's and ten's digits must equal zero. Thus, $h + u - (a + u - 1) = 0$ or $h - a + 1 = 0$ and $h = a - 1$. That is to say, the hundred's digit is one less than the thousand's digit. It follows that $a + h + t + u = a + a - 1 + u - 1 + u = 2a + 2u - 2 = 2(a + u - 1)$

is exactly divisible by 9. Hence, $a + u - 1 = 9$ and $a + u = 10$. Since $h = a - 1$, then $a = h + 1$ and $a + u = h + 1 + u = 10$ or $h + u = 9$. Thus, the sum of the hundred's and unit's digits must equal 9. It follows that the sum of the thousand's and ten's digits must also equal 9, for the difference between these sums is zero. Therefore, if you name the unit's digit, I can reveal the other three digits, for the ten's digit is one less than the unit's digit, the hundred's digit is the difference between 9 and the unit's digit, and the thousand's digit is the difference between 9 and the ten's digit.

31. guessing the total thrown with two dice

Turning around and before picking up the dice, I total the numbers on their faces. In this case, the total is $4 + 6 = 10$. I add 7 to this sum and obtain $10 + 7 = 17$, which is the total sum obtained by you.

Explanation. A die is marked with spots from one to six arranged so that the spots on the opposite faces total seven. Let m and n denote the numbers that turn up in a throw of two dice and suppose you pick up the die that turns up n. The number on the opposite face is $7 - n$ and the sum of the original throw $m + n$ plus the number on the opposite face of the die that you picked up is $m + n + (7 - n) = m + 7$. That is to say, this sum is equal to the number m on the die that has not been picked up plus 7. Suppose that when you again throw the die you picked up, it turns up d. You add d to the sum $m + 7$ obtaining as your final sum $m + d + 7$. When I turn around and before I pick up the dice, I note that the final throw is $m + d$. Hence, to obtain the final sum $m + d + 7$, I simply add 7 to the final throw $m + d$.

32. guessing the throw with two dice #1

You throw 3 and 5 forming the number 35 (or 53). Suppose you select 3; then the successive operations yield $[(3 \times 5) + 7] \times 2 + 5 = 49$. Subtract 14 from this final sum 49; that is $49 - 14 = 35$, to obtain the original throw 3 and 5. If you had selected 5, the final sum is $[(5 \times 5) + 7] \times 2 + 3 = 67$; subtract $67 - 14$ to obtain 53, which is the original throw 5 and 3.

Explanation. Let n and m denote the numbers that turn up on the throw. This throw forms the number $10n + m$ (or $10m + n$). Suppose you select n; then the successive operations yield $(5n + 7) \times 2 = 10n + 14$ and, finally, $10n + 14 + m$, or $10n + m + 14$. If I subtract 14 from this sum, I obtain $10n + m$ which is the number formed by the original throw. Therefore, the original throw is n and m.

33. guessing the throw with two dice #2

You tell me the result $5423 \div 11 = 493$. I subtract 7 and divide by 9; thus $(493 - 7) \div 9 = 54$. Then the original throw was 5 and 4, and the total thrown $5 + 4 = 9$.

Explanation. Let a and b denote the numbers that turn up on the throw of the two dice in order from left to right. This throw forms the number $10a + b$. A die is marked with spots from one to six arranged so that the spots on the opposite faces total seven. Hence, the number on the face opposite a is $7 - a$ and that on the face opposite b is $7 - b$, so that the four-digit number you obtain is $a10^3 + b10^2 + (7 - a)10 + (7 - b) = a10^3 + b10^2 - a10 - b + 77$, or $a(10)(10^2 - 1) + b(10^2 - 1) + 77 = a(10)(99) + b(99) + 77$. Now 99 and 77 are exactly divisible by 11; hence, the number is exactly divisible by 11. Upon dividing by 11, the result is $a(10)(9) + b(9) + 7$. If I now subtract 7, I obtain $a(10)(9) + b(9)$. Dividing by 9 yields $10a + b$, which is the number formed by the original throw. Therefore, the original throw was $a + b$ and the order of the throw a and b.

34. guessing the total thrown with three dice #1

Turning around and before picking up the dice, I total the numbers on their faces. In this case $4 + 4 + 5 = 13$. I add 7 to this sum to obtain $13 + 7 = 20$, which is the total sum obtained by you.

Explanation. Let a, b, and c denote the numbers that turn up in a throw of three dice and suppose you pick up the die that turned up b. The number on the opposite face of this die is $7 - b$ (see problem 31), and the sum of the original throw $a + b + c$ plus $7 - b$ is $a + b + c + (7 - b) = a + c + 7$. Suppose that when you again throw, the number n comes up. Add n to $a + c + 7$, obtaining as your final sum $a + c + n + 7$. When I turn around and before I pick up the dice, I note the final throw $a + c + n$. Hence, to obtain the final sum $a + c + n + 7$, I simply add 7 to your final throw $a + c + n$.

35. guessing the total thrown with three dice #2

Turning around and before picking up the dice, I note the final throw, which is $3 + 6 + 1 = 10$. I add 14 to this sum to obtain $10 + 14 = 24$, which is your total sum.

Explanation. Let a, b, and c denote the numbers that turn up in a throw of three dice. Suppose you pick up the dice that turned up b and c. The numbers on the opposite faces of these dice are $7 - b$ and $7 - c$, respectively. (See problem 31.) Hence, the sum of the original throw $a + b + c$ and the numbers on the opposite faces of the dice picked up is $a + b + c + (7 - b) + (7 - c) = a + 14$. Suppose that when you again throw the two dice, they come up m and n. Now, add m and n to $a + 14$ obtaining as your final sum $a + m + n + 14$. When I turn around and before I pick up the dice, I note the final throw $a + m + n$. Hence, to obtain the final sum, $a + m + n + 14$, I simply add 14 to the final throw $a + m + n$.

36. guessing the throw with three dice #1

You throw 2, 6, and 4 forming the numbers 264 or 426, or any of five other possible combinations of these three digits. Now, suppose you select the 4. The successive operations yield $[(4 \times 2 + 1)] \times 5 = 45$. Suppose you now select and add 2, then multiply by 10, add 7 and finally add the remaining number 6; thus, $(45 + 2) \times 10 + 7 + 6 = 483$. Subtract 57 from this result to obtain $483 - 57 = 426$, which is one of the possible arrangements of the numbers thrown. Therefore, the original throw is 4, 2, and 6.

Explanation. Let a, b, and c denote the numbers that turn up on the throw of three dice, forming the number $100a + 10b + c$ or any of the other five possible combinations of the three digits a, b, and c. Suppose you select c and then a, the successive operations yield,

$$[(2c + 1)5 + a]10 + 7 + b = 100c + 10a + \mathrm{b} + 57.$$

If I now subtract 57 from this result, I obtain $100c + 10a + b$, which is one of the possible numbers formed with the digits a, b, and c originally thrown. Therefore, the original throw is c, a, and b.

37. guessing the throw with three dice #2

You throw 2, 4, and 5. This throw forms the number 245; the corresponding digits on the opposite faces are 5, 3, 2, respectively. (See problem 31.) Hence, you write 245532, divide this number by 37, and then by 3; thus, $(245532 \div 37) \div 3 = 6636 \div 3 = 2212$. Tell me this result and I shall then subtract 7 and divide by 9; thus $(2212 - 7) \div 9 = 2205 \div 9 = 245$, which is the number formed by the original throw. Therefore, the original throw is 2, 4, and 5.

Explanation. Let a, b, and c denote the numbers that turn up on the throw of three dice. Then this throw forms the number $100a + 10b + c$. The corresponding numbers on the opposite faces of a, b, and c are $7 - a$, $7 - b$, and $7 - c$, respectively. Thus, a six-digit number is obtained, which can be written,

$$a10^5 + b10^4 + c10^3 + (7 - a)\,10^2 + (7 - b)10 + (7 - c) = a10^5 + b10^4 + c10^3 - a10^2 - b10 - c + 777.$$

Collecting terms, we may write,

$$a10^2(10^3 - 1) + b10(10^3 - 1) + c(10^3 - 1) + 777$$
$$= a10^2(999) + b10(999) + c(999) + 777$$

Now 999 and 777 are both divisible by 37; hence, the number is divisible by 37.

Dividing by 37 yields $a10^2(27) + b10(27) + c(27) + 21$; then dividing by 3, I obtain $a10^2(9) + b10(9) + c(9) + 7$. This was the number you gave me. If I now subtract 7, I get $a10^2(9) + b10(9) + c(9)$.

Dividing by 9, I obtain $a10^2 + b10 + c = 100a + 10b + c$; this is the number formed by the original throw. Therefore, the original throw is a, b, and c.

38. guessing the total in a column of three dice

The final sum obtained by you is 20. When I turn around, I see the top face of the uppermost die. This top face must be the face opposite 6; that is, $7 - 6 = 1$. Subtract this number 1 on the top face from 21 to obtain $21 - 1 = 20$, which is the final sum.

Explanation. Let a denote the number on the face of the first die that touches the table. Then the opposite face of this die is $7 - a$. (See problem 31.) Suppose b represents the number on the face of the second die that touches the first die, then the sum $a + (7 - a) + b$ is the sum obtained in step 3. The opposite face of the second die must be $7 - b$. Suppose c denotes the number on the face of the third die that touches the second. Adding the sum $(7 - b) + c$ to $a + (7 - a) + b$, yields $a + (7 - a) + b + (7 - b) + c = 14 + c$, which is the sum obtained by you. When I turn around, I see the top face of the third die, which must be $7 - c$.

If I then subtract $7 - c$ from 21, I obtain $21 - (7 - c) = 14 + c$, which is the sum you obtained.

39. guessing the hour #1

The hour you touch is VIII. I ask you to count in a counterclockwise direction 12 more taps than the hour you touch. Since you touched VIII, I ask you to count $8 + 12 = 20$ taps beginning to count to yourself with the number of the hour you thought of; thus, if the hour you thought of is IV, count 4, 5, 6, etc. to 20. Upon reaching the count of 20, you will be at the IV.

Explanation. Let a denote the hour you thought of and b the hour you touched. Since you count in a counterclockwise direction, adding 12 to the hour touched; that is, b, you will always arrive at XII. If you now count a taps from the XII in a clockwise direction, you will arrive at the hour you thought of. Hence, you will get to the hour you thought of at the $(b + 12 - a)$th tap. If I ask you to begin counting with a, then $a + 1$, etc., you will tap $b + 12 - a + a = b + 12$ times to get to a.

40. guessing the hour #2

I ask you to count in a counterclockwise direction 13 more taps than the hour I name. Since I named the IX, I ask you to count $9 + 13 = 22$ taps to yourself from the hour you thought of. Thus, if you thought of the III, you count 4, 5, 6, etc. until you reach 22. You will then be at the III, the hour you thought of.

Explanation. Let a denote the hour you thought of and b the hour I name. Since you count in a counterclockwise direction adding 12 to the hour b named by me; you will always arrive at the XII. If you now count a

taps from XII in a clockwise direction, you will arrive at the hour you thought of. Hence, you will reach the hour you thought of at the $(b + 12 - a)$th tap. Thus, if I ask you to begin counting at $a + 1$, $a + 2$, etc., you will tap $b + 12 - a + a + 1 = b + 13$ times to get to a.

41. guessing the card

Select one of the first 10 cards and, by dealing, I transfer 15 cards from the bottom to the top of the card. You begin counting the top card as 8, the next one 9, etc. I let you count to $10 + 15 + 1 = 26$. This card is the 8 of diamonds.

Explanation. There are 52 cards in the deck and suppose you select the xth card out of the first n cards. By dealing, I reverse the order of the first n cards and transfer p cards, where p is less than $52 - n$, from the bottom to the top of the deck. This makes the card originally selected the $(p + n - x + 1)$th card from the top. Hence, if I ask you to count the top card as $x + 1$, then the next card $x + 2$, etc., the card originally selected is the $(p + n + 1)$th card.

Note that we can change at will the number of the first n cards and the number of cards p transferred from the bottom to the top of the deck. For example, if I ask you to choose from the first 15 cards and transfer 20 cards from the bottom to the top of the deck, the card you select will be the $15 + 20 + 1 = 36$th card beginning the count with one more than the number indicating the original position of the card you selected.

42. guessing the total sum of cards

I turn around and count the number of piles, in this case 5. You have given me 2 cards left over. The sum of the numbers on the bottom cards is $13(p-4)+r$, where p is the number of piles and r the number of cards left over. Here $p=5$ and $r=2$. Therefore, the sum of the numbers on the bottom cards is $13(5-4)+2=15$.

Explanation. If n_1 is the number on the bottom card of the first pile, then the number of cards in that pile is $13-n_1$. Similarly, the number of cards in the second pile is $13-n_2$, where n_2 is the number of the card on the bottom of the second pile, etc. If there are p piles, then the number of cards in all the piles is the sum of the p terms; thus,

$$(13-n_1)+(13-n_2)+\ldots+(13-n_p).$$

Since there are 52 cards in a deck, if we let r denote the number of cards left over, then,

$$(13-n_1)+(13-n_2)+\ldots+(13-n_p)+r=52,$$

or $$13p-(n_1+n_2+\ldots n_p)+r=52,$$

and $$13p-52+r=n_1+n_2+\ldots+n_p,$$

so that $$13(p-4)+r=n_1+n_2+\ldots+n_p.$$

The right-hand side of this equation is the sum of the numbers on the bottom cards. Therefore, $13(p-4)+r$ is the sum of the numbers on the bottom cards, recalling that p is the number of piles and r the number of cards left over.

43. revealing a card #1

You tell me the card you selected is in the third row of the lower array. I shall immediately tell you that it is the third card in the column headed by the 7 of diamonds, which is the 8 of diamonds.

Explanation. If I pick up the cards in each column in order from left to right and then deal them out in rows of four cards each beginning at the top left-hand corner, I have simply changed the rows of the first array to columns, forming the lower array. It is easily seen that if I remember the card in the first column of a particular row in the upper array, it will tell me the column in which the card you have selected is located in the lower array. If you now tell me the number of the row in which the card you selected is located in the lower array, I know the column and the row in which the card is located, therefore I can immediately pick it up.

44. revealing a pair of cards

M	U	T	U	S	N	O	M	E	N
♡9	♡Q	♠3	◇Q	♣9	◇5	♡7	♣5	♠K	♠J
D	E	D	I	T	C	O	C	I	S
♣3	◇3	♡K	◇9	◇6	♡2	♠6	♣4	♠7	♡8

Imagine the magic words written on a table so that each word forms a line, as shown above. This sentence contains four words of five letters each and each letter appears twice only. I take up the pairs in any order whatever, making sure that each pair is kept together. Suppose I take up the pairs in the following order:

♡2 – ♣4, ♡7 – ♠6, ◇5 – ♠J, ◇9 – ♠7,

◇3 – ♠K, ♣3 – ♡K, ♣9 – ♡8, ♠3 – ◇6,

♡Q – ◇Q, and ♡9 – ♣5.

The top card in the pile is the 9 of hearts. I deal this card where the M in MUTUS would be. The next card which is the pair of the first, the 5 of clubs, is placed on the second M, which is the M in NOMEN and is the third card in the third row. The third card, in this case, the queen of hearts, is placed on the first U in MUTUS, which is the second card in the first row. The fourth card, which is the pair of the third card, the queen of diamonds, is placed on the second U in MUTUS, which is the fourth card in the first row. The next two cards which form a pair, the 3 of spades and 6 of diamonds, are placed on the two T's, and so on as shown in the figure. As soon as I am told the row (or rows) in which the two cards are located, I shall find them on the letters common to the words in the row (or rows). Thus, if I am told the cards are in the second and third rows, they must correspond to the two E's; hence, they are the 3 of diamonds and the king of spades.

45. revealing a card #2

The card you selected is the middle, or fifth, card in the last pile you indicate.

Explanation. I deal the cards in three piles of nine cards each. The pile containing the card you selected is taken up so that it is between the other two. Thus, at the top of the pack there are nine cards, next there are nine cards containing the selected card, and finally nine more cards. When the cards are dealt a second time, the bottom three cards in each pile will be from the first pile taken up, the next three cards from the pile containing your choice, and the remaining three cards from the third pile. Now I take up the piles again so that the one containing the selected card is between the other two piles. Then, at the top of the pack there are nine cards, next three cards, then three cards containing the selected card, and, finally, the remaining 12 cards. I deal the cards for the third time. In each pile the bottom three cards will be from the first nine cards at the top of the pack, the next, or fourth, card from the following three cards, and the fifth card from the three cards containing the selected card. That is to say, the card you selected is the fifth card in one of the three piles. If you now tell me the pile in which the card is located, I can tell you the card you selected.

46. the christians and turks

The captain arranged the passengers as follows:

CCCCTTTTTCCTCCCTCTTCCTTTCTTCCT,

where C denotes a Christian and T a Turk. This order may be remembered by the position of the vowels in the following sentence,

Populeam jirgam Mater Reginia ferebat,

where *a* stands for 1, *e* for 2, *i* for 3, *o* for 4 and *u* for 5. Then the order is *o* Christians, *u* Turks, *e* Christians, and so on.

47. revealing a selected digit and the age of a person

Let $n =$ the digit selected and let $10h + u$ be your age. Then, if the current year is 1969, the year in which you were born is $1969 - (10h + u) = 1969 - 10h - u$. The operations to be performed are,

$$[(2n + 5) \times 50] + 1719 - (1969 - 10h - u)$$

But, this equals,

$$100n + 250 + 1719 - 1969 + 10h + u$$

The expression, $100n + h$ represents a number whose hundred's digit is n (the selected digit) and whose last two digits form the number $10h + u$, which is your age.

chapter 5
teasers for the thinker

1. the wise tourist

Before the great flood, there existed a land inhabited by two tribes, the Alphas and the Omegas. The Alphas were such liars that they *never* told the truth and the Omegas were so honest that they *always* told the truth. One day, a tourist arrived from a distant land, and he wanted a guide to take him around and show him the country. Knowing the reputation of the inhabitants, he wanted to hire an Omega as a guide. He approached a group of three men and asked the first whether he was an Alpha or an Omega. This man, who spoke a dialect, said something to the tourist which he could not understand. The second man then said to the tourist, "He said that he is an Omega. He is an Omega, and so am I." However, the third man promptly stated, "That's a lie," and pointing to the first man, he said, "He is an Alpha and I am an Omega." The tourist without hesitation hired one of the men, who was an Omega. How did he know which one of the men to hire?

2. the quick-thinking executive

Joe, Jack, and Jim are executives of a large corporation. All three of these men are very intelligent. The president of the corporation wanted to advance one of them to executive vice-president; so he decided to give the job to the one who showed the greatest ability for quick, logical thinking. To find the quickest thinker among them, he had all three men seated in a room without mirrors so that they were facing each other. Then he said to them, "I am going to put the lights out; then I shall mark each of your foreheads with either a black or a white spot. After doing this, I shall turn the lights back on, and if any one of you sees a black mark on the forehead of one of the others, stand and remain standing until one of you can let me know the color of the mark on your own forehead. The first one who can do this will be appointed executive vice-president."

Accordingly, the president turned the lights out and marked each man's forehead. Then he turned the lights on again and all three of them stood up. Immediately, Joe said, "I have a black mark." How did he know this?

3. the commuter train

A commuter train from New York City to White Plains, N. Y., left Grand Central Station in New York City and made its first stop at the 125th street station where 20 passengers got aboard. At the next stop, Mount Vernon, half of the passengers got off and 20 new passengers boarded the train. At Bronxville, a third of the passengers got off and eight new ones boarded. At Tuckahoe, a quarter of the passengers got off and nine new ones got on. At Scarsdale, a fifth of the passengers left the train and six new ones got aboard. At Hartsdale, three tenths of the passengers got off and only one boarded. Finally, at White Plains, the last stop, 106 passengers got off the train. How many passengers were on the train when it left Grand Central Station?

4. the counterfeit coin

A bank teller knows that one out of every eight coins is counterfeit and that the counterfeit coin will weigh less than a good coin. He has a balance (with two pans) to compare the weights of the coins. How can he identify the counterfeit coin with only two weighings?

5. the birthday party

At a birthday party for one of my grandchildren, Tony and Chris wanted to sit next to Cookie. For the sake of peace and order, Paul should not sit next to Tony or Chris. Peggy does not want to sit next to Tony or Chris. Tory, whose actual name is Victoria, and Jeffer will sit opposite each other. Bill wants to sit next to Cathy, but Cathy wants to sit next to Jeffer. Finally, Julia does not want to sit next to Bill. While these arrangements are being discussed, Paul and Cathy are already seated and are beginning to help themselves to the ice cream and cake. I want to arrange the seating so that the boys and girls are alternately seated. Luckily, the seats taken by Paul and Cathy, on the round table, allow me to seat them all in a satisfactory arrangement. How were they seated?

6. the dinner party

Mr. and Mrs. Adams were celebrating Mr. Adams' birthday with a dinner party. They invited five married couples and Mrs. Adams arranged the seats at her round dining room table so that the men and women sat alternately, no husband sat next to his wife, and each husband was separated from his wife by the same number of places. The following facts were observed:

1) Mrs. Burke was seated opposite Mrs. Adams.
2) Mr. Clark was three places to the right of Mrs. Adams.
3) Mr. Dexter was three places to the left of Mrs. Clark.

4) Mrs. Egan was two places from Mrs. Adams.
5) Mrs. Dexter was opposite Mrs. Foster.

How were the twelve people seated?

7. el camino del diablo

El Camino del Diablo (The Devil's Road) extended from the town of Sonoita, on the border of Arizona and Mexico, to the mouth of the Gila River, where the city of Yuma now stands. This road acquired its name during the gold rush days. It passed through a barren, waterless desert and was about 135 miles long. An expedition was organized to travel on foot 50 miles of this road starting from Sonoita and returning there. The men could walk 20 miles a day, but they could carry only enough food and water for two days, and each man had to consume one day's rations each day. In order to make the round trip without outside help, the expedition cached provisions along the road. What is the shortest time needed for the expedition to deposit the caches of food and make the round trip?

8. a hat trick

This old puzzle sometimes takes the form of a fairy tale. Once upon a time, the king of Utopia decided to give his beautiful daughter in marriage to the most capable young man in his kingdom. Four young men, Alert, Wise, Clever, and Shrewd, where finally selected from a very large number of candidates as being the most gifted and intelligent. In order to make the final decision, the king devised a fair test.

The four candidates were blindfolded and seated around a table. Then the king said to them, "I shall place a black or a white hat on each of your heads. Then I shall have your blindfolds removed simultaneously. If any of you sees more white hats than black ones, raise your right hand, and, as soon as you can tell me the color of the hat you are wearing, stand up and give me a convincing argument for your decision. The first man who can tell me the color of his hat and the reason for his choice may then marry my daughter."

The king then placed a hat on each man's head and had the blindfolds removed simultaneously. All four men raised their right hands and almost immediately Wise stood up and said, "Your Majesty, I have a white hat." How did he justify his statement?

9. a hat trick for the ladies

In a girls' college dormitory, a question arose as to how intelligent Pat was. In order to test her reasoning power, the house mother, who taught logic, suggested a test: Two other girls, Carol and Ann, who were also very intelligent, were chosen. The house mother sat them so that Ann saw Carol and Pat, Carol saw Pat only, and Pat saw none of the others. She then took five skull-caps, three green and two black, and placed a cap on the head of each girl while hiding the others. She then asked Ann and Carol the color of their caps, but neither of then could tell her. However, Pat, without seeing the other two girls, stated the color of her cap. What was the color of Pat's cap and how did she arrive at her conclusion?

10. the yard master's problem

A railway track DEF (in the figure) in a railroad yard has two shunting tracks DBA and FCA connected at A. The portion of the tracks at A, common to both

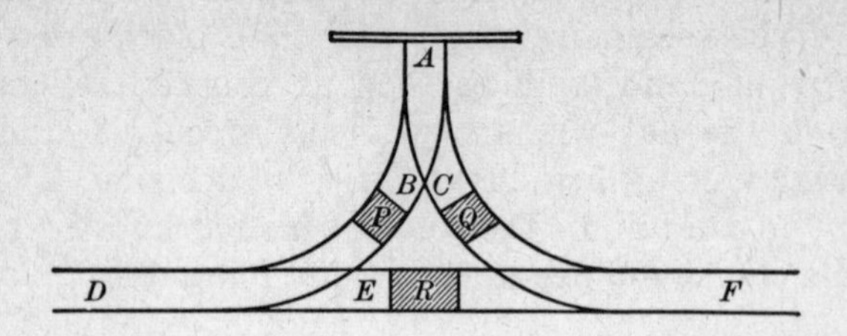

shunting tracks, is long enough to hold only a single car, such as P or Q, but not long enough to contain a whole engine R. Hence, if an engine runs up one of the shunting tracks, say DBA, it has to come back the same way. If car P is at B, car Q is at C, and an engine R, longer than either P or Q, is at E. How can the yard master interchange the cars P and Q?

11. the master detective

Martin, the informer, was murdered. Ricky the Rock, Spanish Joe, and Harry the Hawk were arrested and questioned by D. Tect, the famous detective.

Ricky the Rock was questioned first.

D. TECT. Did you kill Martin?
RICKY. No, I did not. (1)
D. TECT. When was the last time you saw Spanish Joe?
RICKY. I have never seen Spanish Joe before. (2)
D. TECT. Did you know Martin?
RICKY. Yes, I knew Martin. (3)

Next, Harry the Hawk was questioned.

D. TECT. Did you kill Martin?
HARRY. No, I did not. (1)
D. TECT. Did Ricky the Rock see Spanish Joe?

HARRY. The Rock lied when he said he had never seen Spanish Joe. (2)

D. TECT. Do you know who killed Martin?

HARRY. No, I do not. (3)

Then Spanish Joe was questioned.

D. TECT. Did you kill Martin?

JOE. No, I did not. (1)

D. TECT. Do you know Ricky the Rock and Harry the Hawk?

JOE. Yes, they are my pals. (2)

D. TECT. Did the Rock kill Martin?

JOE. No, the Rock has never killed anybody (3)

Now, D. Tect knows that one, and only one, of each man's statements is false, and that one of the three men killed the informer. How did he conclude, from these statements only, who killed Martin?

12. d. tect gets his man

In an assault and robbery case there were five witnesses, Brown, Cohen, O'Neil, Lash, and Smith. In an effort to obtain a description of the attacker, these witnesses were asked to give the following information.

		STATEMENTS		
Witness	Age (1)	Color of hair (2)	Color of eyes (3)	Color of jacket (4)
Brown	40	brown	blue	dark grey
Cohen	30	black	brown	black
O'Neil	40	brown	blue	dark brown
Lash	30	blond	brown	dark blue
Smith	25	brown	brown	dark blue

From these descriptions, the famous detective, D. Tect, discovered the guilty man. For future cases, he compared the actual likeness of the man with the descriptions given by these witnesses. He found that each witness had made three incorrect statements, but that each of the four questions had been answered correctly at least once. What did the guilty man actually look like?

13. murder in the back room

Four men were playing poker in the back room of a tavern. Suddenly, a shot was heard. The proprietor of the tavern rushed into the back room; he found one man dead on the floor and the three other men standing around the dead man. A gun was lying on the floor next to the dead man. The dead man's name was Sharp and the other three men were Lamb, Clod, and Dolt. These three were arrested and, when questioned at the police station, they made the following statements:

LAMB. I didn't kill him. (1)
Sharp did not commit suicide. (2)
I was a friend of Sharp. (3)

CLOD. Sharp did not commit suicide. (1)
I was sitting across the table from Sharp. (2)
Lamb was not a friend of Sharp. (3)

DOLT. Clod did not kill Sharp. (1)
Sharp committed suicide. (2)
Lamb lied when he said he was a friend of Sharp. (3)

D. Tect, the famous detective, knew that one, and only one, of each man's statements was false. How did he discover, from these statements only, who killed Sharp?

14. d. tect cried murder

The policeman on duty in Hell's Acres came to the scene of the crime and found a clergyman, an ex-convict on parole, and an electrician. There was no doubt that one of these men had committed the crime. The names of these men were Divine, Knave, and Cell, but not necessarily in that order. On being questioned by the famous detective, D. Tect, each man made the following statements:

KNAVE. Divine didn't do it. (1)
Cell did it. (2)
CELL. I didn't do it. (1)
Divine did it. (2)
DIVINE. I didn't do it. (1)
Knave did it. (2)

Upon investigating the crime, D. Tect found that the two statements made by the clergyman were true, both statements made by the criminal were false, and the ex-convict had made one true and one false statement. What are the names of the clergyman, the ex-convict, and the electrician? Who is the criminal?

15. who's who?

Clark, Morris, and Howe are a physicist, a pianist, and a philosopher, but not necessarily in that order. Determine who's who by the following facts:

1) Morris makes more money then Clark.
2) Clark has never heard of Morris.

3) The philosopher tried to get the pianist to do a benefit for his charity, but the pianist refused, because he was doing some experiments on sound with the physicist.
4) The physicist makes more money than the philosopher.

16. the bridge game

Bill, Sam, and Jack are married to Grace, Joan, and Becky, but not necessarily in that order. Last Friday, they played bridge as usual, but no wife was a partner of her husband. In one of the rubbers,

1) Bill and Grace play Jack and Becky.
2) Joan plays with Jack as a partner.
3) Bill plays with Sam's wife as a partner.

What is the name of each man's wife?

17. who verifies the accounts?

Cronin, Baker, and Ennis are the controller, manager, and treasurer of a corporation, but not necessarily in that order.

1) The treasurer, who is single, earns the least.
2) Ennis is a brother-in-law of Baker.
3) Ennis earns more money than the manager.

Who is the controller?

18. untie this nuptial knot

Dan, Pete, and Sid are married to Cora, Jill, and Flo, but not necessarily in this order, and they are all members of the same golf club. In the club's golf tournament,

1) Dan's wife and Jill's husband play golf against Cora and Flo's husband.
2) No husband and wife play as partners.
3) Pete, who is a tennis player, does not play golf.

What is the name of each man's wife?

19. the train puzzle

A train is operated by three men, Jones, Robinson, and Smith, who are engineer, fireman, and brakeman, but not necessarily in that order. On the train, there are three passengers, Mr. Jones, Mr. Robinson, and Mr. Smith.

1) Mr. Robinson lives in Detroit.
2) Mr. Jones earns exactly $20,000 yearly.
3) The brakeman lives halfway between Chicago and Detroit.
4) Smith beat the fireman at billiards.
5) The brakeman's next door neighbor, one of the passengers, earns exactly three times as much as the brakeman.
6) The passenger whose name is the same as the brakeman's lives in Chicago.

What are the names of the engineer, fireman, and brakeman?

20. it's a wise son who knows his own father

Al, Fred, and Ed are married to Alice, Grace, and Rose, and they have one child each, Mary, Tony, and Jane, but not necessarily in that order.

1) The children of Ed and Grace are on the swimming team that competes against the Riviera School for girls.
2) Jane is not Al's daughter
3) Rose is not Fred's wife.

Who belongs to which family?

21. the professionals

Henry, Bob, Maggie, and Ann are a doctor, a lawyer, an architect, and a draftsman, but not necessarily in this order.

1) The doctor, who is married, earns more than his wife, the architect.
2) Ann has never heard of the vanishing point in perspective drawing.
3) Bob is a confirmed bachelor.

What is the profession of each of these persons?

22. the executive officers

The president, vice-president, treasurer, and secretary of the Great Rivers Co. are Mr. Blair, Mr. Koch, Mr. Lewis, and Mr. Warren, but not necessarily in that order.

1) Mr. Blair and the president were in the same unit in the army.
2) Mr. Koch will probably be the next president.
3) Mr. Lewis and the secretary play golf together.
4) Mr. Blair beats the treasurer at high-low jack.
5) Mr. Lewis served in the navy.
6) Mr. Koch does not play golf or cards.

Who are the executive officers?

23. whose little girl are you?

During the father-daughter weekend at a women's college, six couples took part in the table tennis competition. The fathers' names were William, Robert, Thomas, Edward, Arthur, and Carl; the daughters' names were Pat, Joan, Dorothy, Nancy, Mary, and Helen, but not necessarily in that order. Their homes were in New Jersey, Virginia, Connecticut, Pennsylvania, Massachusetts, and New York.

1) Robert and Edward played doubles against Joan and Mary.
2) The girl from Pennsylvania played singles against Dorothy, Pat, Helen, and Thomas.
3) William and Carl played doubles against Pat and Mary.
4) The girl from Connecticut played singles against Mary, Dorothy, Pat, and Carl.
5) Mary and Joan played doubles against Robert and William, and then against Carl and Edward.
6) Dorothy played singles against Nancy and Helen.
7) The girl from Virginia played singles against Dorothy.
8) Joan played singles against William, Carl, and Thomas.
9) The girl from New Jersey played singles against Robert and William.
10) Edward played singles against Pat and Nancy.
11) Pat played singles against Dorothy, Helen, and Nancy.
12) William played singles against Helen and Nancy.
13) Nancy played singles against the girls from Pennsylvania and New Jersey.
14) Nancy played singles against Pat and Helen.

15) The girl from Massachusetts played singles against Dorothy and Nancy.

No father and daughter ever took part in the same match. Find the father-daughter relationship and their respective home states.

24. the traveling salesmen

Six salesmen met in the club car of the Trans Continental Express. They sat in two rows facing each other, three on each side. Their names are Stark, Hart, Duffy, Carter, Waters, and Ring; they sell carpets, toys, glassware, hardware, boats, and jewelry, but not necessarily in that order. They are reading the catalogs published by their companies, but no salesman is reading his own company's catalog.

1) Stark is reading about hardware.
2) Duffy reads the catalog published by the company whose salesman is sitting just opposite him.
3) Hart sits between the hardware and jewelry salesmen.
4) The hardware salesman sits opposite the toy salesman.
5) Carter reads the carpet salesman's catalog.
6) Hart is a relative of the glassware salesman.
7) Waters sits next to the carpet salesman.
8) Stark sits at the end of a row, and he is not interested in toys.
9) Carter sits opposite the glassware salesman.
10) Ring lives in the south, and he is not interested in carpets.

What product does each man sell?

25. the performing arts

Hall, Reno, Marinelli, and Pastore are a dancer, a painter, a writer, and a pianist, but not necessarily in that order.

1) Marinelli and Hall were in the audience the night the pianist gave a recital.
2) Both Pastore and the writer have sat for portraits by the painter.
3) The writer, whose biography of Reno was a best seller, is planning to write a biography of Marinelli.
4) Marinelli has never heard of Hall.

What is each man's profession?

26. nepotism

The offices of president, vice-president, secretary and treasurer of a corporation were held by members of the same family. A very good customer from South America was visiting the president of this corporation.

1) One night, the customer had dinner with the following office holders: the president, the treasurer, Mr. Burr, and Mr. Fox.
2) Mr. McCue, who holds one of the offices, is the president's cousin.
3) Mr. Burr's wife is the secretary's sister.
4) Mr. Fox is married to the president's sister.
5) Since Mr. Marx was not present at the dinner, the customer was invited again and Mr. Palmer, Mr. Fox, and Mr. Marx had dinner with him.

Which man holds which office?

27. sonny boy's letter

It seems that, once upon a time, a college student found himself with some debts, a heavy date coming up, and no money. His credit among his fellow students was not too good and, since it was the end of the month, all his friends were also short of funds. As a last resort, he decided to appeal to his father. Having a sense of humor and knowing his father was an expert at deciphering codes, he sent his father the following message.

Dear Dad,

Each letter in this message represents a different digit. Please send me the amount of cents shown by this sum.

```
   S E N D
 + M O R E
 ---------
 M O N E Y
```

Love to Mom.

Signed,
Sonny Boy

How much money did Sonny Boy ask for?

28. dad's answer to sonny boy's letter

The story goes on to tell that Sonny Boy's father deciphered his code message, but, not wishing to spoil the young man, he decided to answer him saying that it was impossible to grant his request; so he sent the following code message,

$$\begin{array}{r} S\ P\ E\ N\ D \\ -\ L\ E\ S\ S \\ \hline M\ O\ N\ E\ Y \end{array}$$

If each letter represents a different digit, show how Dad refused Sonny Boy's request?

29. dad thinks it over and writes again

After answering Sonny's letter and refusing to send any money, Dad thought it over. He decided that, since Sonny Boy was getting good marks in all his subjects, he would send him the money. However, by that time, he had misplaced Sonny's letter, so when he sent the following message, he made a mistake of transposition in the amount.

Dear Son,

I have though it over and I am sending you the largest amount of cents represented by this sum.

```
   S A V E
 + M O R E
 ---------
 M O N E Y
```

Love,
Dad

If each letter represents a different digit, how much money did Dad send to Sonny Boy?

30. sonny boy goes on a business trip

When Sonny Boy finished college, he went to work for dear old Dad. While Dad was away, Sonny Boy was asked to go on a business trip to buy some raw material for their factory. To make sure they could take advantage of a good offer, Dad sent the following telegram indicating the amount he could spend. As in No. 29, if each letter represents a different digit, how much could Sonny Boy spend?

```
   T A K E
 + M O R E
 ---------
 M O N E Y
```

31. a little business advice

Dad got back from his trip to hear that Sonny Boy needed more money to close a business deal. To let him know how much more he could spend and, at the same time, to give a little advice, Dad sent this telegram.

$$\begin{array}{r} D\ E\ F\ E\ R \\ -\ D\ U\ T\ Y \\ \hline N\ O\ G\ O \end{array}$$

If each letter represents a different digit, how much more money could Sonny Boy spend?

32. things are not what they seem

The following subtraction is obviously incorrect. However, if each letter represents a different digit, we can obtain various correct solutions.

$$\begin{array}{r} F\ O\ U\ R \\ -\ T\ W\ O \\ \hline T\ E\ N \end{array}$$

Obtain the solution that gives any one of these letters its greatest possible value.

33. deceived by appearances

As in the preceding cryptogram, the operation as stated seems to be incorrect. However, if each letter represents a different digit, we can obtain various correct solutions.

```
  T H R E E
 - F O U R
 ---------
   F I V E
```

Obtain a solution that gives each letter its greatest possible value.

34. the cat

In the following long division example, each letter represents a different digit.

```
          C A T
     ----------
R C ) A P D T M
      A D C
      -----
        N N T
          B A
        -----
          A E M
          A E M
          -----
```

Show that this division is correct.

35. the arab

As in the preceding example, each letter represents a different digit.

```
              A R A B
        ______________
  P R E)A N M M R D R
          P R E
        _______
          C S R R
          C C P R
          _______
            A E M D
              P R E
            _______
              D E A R
              D E A R
              _______
```

Show that this long division example is correct.

36. a good one

An easy long division example is represented by the following cryptogram in which each letter stands for a different digit.

```
              G O O D O N E
        ____________________
  P R R)D R E O O O D M N
        D R M
        _____
          P O O O
            R R D
            _____
              D D M
              G R E
              _____
              P D R N
              P D R N
              _______
```

Find the correct solution.

37. the three fives

The answer to the cryptogram in problem 32 became so blurred that all that was left were the few figures shown below,

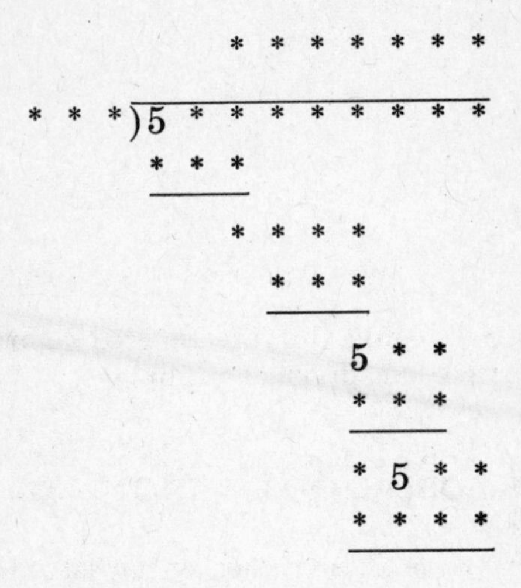

The asterisks, of course, stand for the blurred figures. Without referring to the cryptogram in No. 32, reconstruct the entire long division example.

38. the five fours

A very old parchment shows a long division example in which all the figures are illegible except five fours, as shown below.

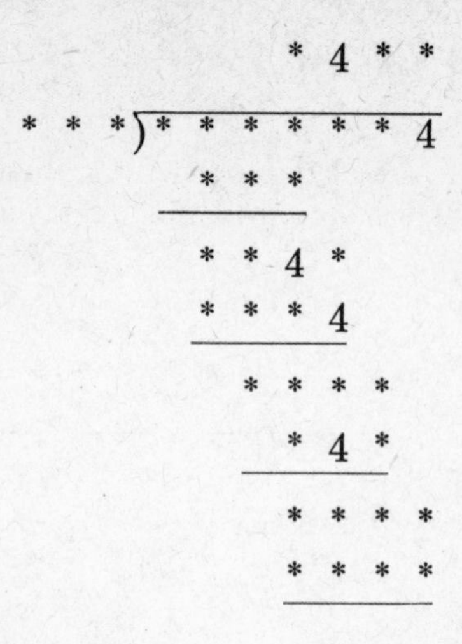

If the asterisks stand for the blurred figures, reconstruct the entire long division example.

39. the suspicious sailors #1

Four sailors were shipwrecked on an island. In order to make sure that they would have enough to eat, they spent the day gathering bananas. They put the bananas into one large pile, which they decided to divide equally among themselves the next day. As it happened, each sailor mistrusted the others; so, after they had all gone to sleep, the first sailor woke up and divided the pile of bananas into four equal parts, but he found that he had one banana left over, which he gave to a monkey. He then hid his share and went back to sleep. Then the second sailor woke up and divided the remaining pile into four equal parts. He also had one banana left over, which he gave to a monkey. Then he hid his share and went back to sleep. Similarly, the third and fourth sailors, in turn, repeated this process; in each case, there is a banana left over, which was given to a monkey. The next morning, when the sailors got up, the pile was without doubt smaller, but, since they all felt guilty, none of them said anything and they agreed to divide the remaining pile into four equal parts. Again, there was a banana left over, which they gave to a monkey. What was the number of bananas in the original pile?

40. the suspicious sailors #2

Five sailors and a monkey were shipwrecked on an island. They spent a day gathering bananas into a pile which they decided to divide equally among themselves the next day. Each sailor mistrusted the others, so, during the night, one of the sailors stole up to the pile of bananas, divided it into five equal parts and took his share; there was one banana left over, so he gave it to the monkey. Each of the other four sailors, in turn, divided the pile of bananas he found into five equal parts without any of the others seeing him, and, in each case, there was an extra banana left over, which was given to the monkey. On the following day, they met and divided the remaining pile into five equal parts and, again, there was one banana left over, which was given to the monkey. How many bananas were in the original pile?

41. the scientific expedition

A scientific expedition set out to cross the Sahara Desert from the north to the south, a distance of 1656 miles. There was no gasoline available along the route that the expedition was to follow. It was, therefore, necessary to store the gasoline at various points along this route, so that the expedition could be made without outside help. A special truck was used that could carry the necessary scientific equipment, as well as 90 gallons of gasoline, including the gasoline in the tank. If this truck was able to use 1 gallon of gasoline for each 12 miles, at which points (in miles) along the route should deposits of gasoline be stored, so that the minimum amount of gasoline would be used for the expedition?

solutions

1. the wise tourist

Although the tourist did not understand what the first man said, he realized that this man had to say, "I am an Omega," for, if he were an Alpha, he would have lied and said that he was an Omega and, if he was an Omega, he would have told the truth and said that he was an Omega. Hence, when the second man stated that the first man had said he was an Omega, he was telling the truth. Therefore, he must be an Omega, and since he said that he and the first man are both Omegas, this must be the truth. It immediately follows the third man lied, so he is an Alpha. Thus, the tourist hired the second man.

2. the quick-thinking executive

While the president was marking the foreheads of each of the men, Joe reasoned as follows: The president cannot mark us all white, for if he did, no one would stand up when the lights are turned on, and we shall immediately know that we are all marked white. He must mark at least two of us black, for if he marks only one of us black, then one man will see no black marks; hence, he will remain seated and immediately realize his mark is black. Moreover, if the president makes only two black marks, then, when we all stand up, one of the two bearing black marks will conclude at once that he has a black mark, for otherwise, since he sees one white mark, there would be less than two black marks and, in that case, not all of us will stand up.

That is to say, if the president marks two men black and one white, the test will not be the same for each man, hence it will not be a fair test. Therefore, all marks on our foreheads must be black. Consequently, as soon as the lights went on, and they all stood up, Joe said, "I have a black mark."

3. the commuter train

Since one passenger boarded at Hartsdale and 106 passengers disembarked at White Plains, the number of passengers left in the train at Hartsdale was $106 - 1 = 105$, but three tenths of the passengers in the train got off at Hartsdale. Hence, there were $1 - \frac{3}{10} = \frac{7}{10}$ of the passengers left in the train at Hartsdale. That is, $\frac{7}{10}$ of the passengers that arrived at Hartsdale was 105, so that the number of passengers that arrived at Hartsdale was $\frac{10}{7}(106 - 1) = \frac{10}{7} \times 105 = 150$. Similarly, the number of passengers that arrived at Scarsdale is $\frac{5}{4}(150 - 6) = \frac{5}{4} \times 144 = 180$. The number of passengers that arrived at Tuckahoe is $\frac{4}{3}(180 - 9) = \frac{4}{3} \times 171 = 228$. The number that arrived at Bronxville is $\frac{3}{2}(228 - 8) = \frac{3}{2} \times 220 = 330$. The number of passengers that arrived at Mount Vernon is $2(330 - 20) = 2 \times 310 = 620$. Finally, the number of passengers that arrived at the 125th street station is $620 - 20 = 600$. Therefore, 600 passengers left Grand Central Station.

4. the counterfeit coin

The teller places three coins on each pan of the balance. If these coins do not balance, the counterfeit coin is one of the three coins in the higher pan. The teller then takes these three coins and proceeds to the second weighing by taking two of these three coins and placing one on each pan of the balance. If these two coins balance, then the third coin is counterfeit. If they do not balance, the counterfeit coin is in the higher pan. Now, suppose the three coins in each pan balance; then the six coins are good. Hence, the counterfeit coin is one of the remaining two. The teller then weighs these two coins to determine which one is good.

5. the birthday party

Assume that Paul sits at seat 1, and the boys and girls sit alternately. Hence the boys may sit on seats 3, 5, 7, and 9 and the girls sit on seats 2, 4, 6, 8, and 10. Since Tony and Chris sit next to Cookie, Cookie must be between Tony and Chris. Moreover, since Bill wants to sit next to Cathy and Cathy wants to sit next to Jeffer, then Cathy must sit between Bill and Jeffer. It follows that neither Cathy nor Cookie may sit on seats 2 or 10, but they may sit on seats 4, 6, or 8. If Cookie sits on seat 6, then Cathy cannot sit on seats 4 or 8, for she sits between Bill and Jeffer, and seats 7 and 5 will both be occupied. Similarly, Cathy cannot sit on seat 6. Hence, Cookie and Cathy may sit either on seats 4 or 8. Assume that Cookie sits on seat 4, then Peggy cannot sit on seats 2 or 6, for she would be next to Tony or Chris and she does not want to sit next to them; hence, she must sit on seat 10. Now Tory cannot sit on seat 6 for she must be opposite Jeffer, so she must sit on seat 2. It follows that Jeffer must sit on seat 7, opposite Tory. Cathy must sit on seat 8, and Bill on seat 9. Julia must then sit on seat 6 and either Tony or Chris on seat 5. If Tony sits on seat 5, then Chris must sit on seat 3. If Cookie sits on seat 8, the reader may check that the relative positions will be the same, but in a counterclockwise direction.

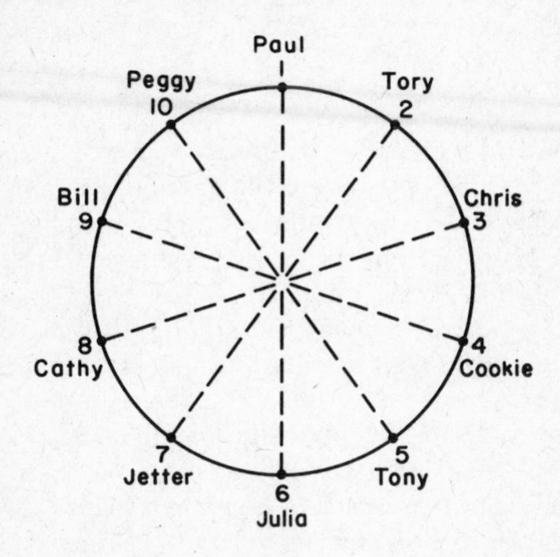

6. the dinner party

Let the capital letters represent the husbands and the lower case letters represent the wives, then Mr. Adams is *A*, Mrs. Adams is *a*, etc. Since, by 1, Mrs. Burke was placed opposite Mrs. Adams, then, assuming Mrs.

Adams was in seat 12, Mrs. Burke was in seat 6. That is, a was in seat 12 and b was in seat 6. By 2, Mr. Clark's place was three places to the right of Mrs. Adams; thus C was in seat 9. We then have the arrangement shown below.

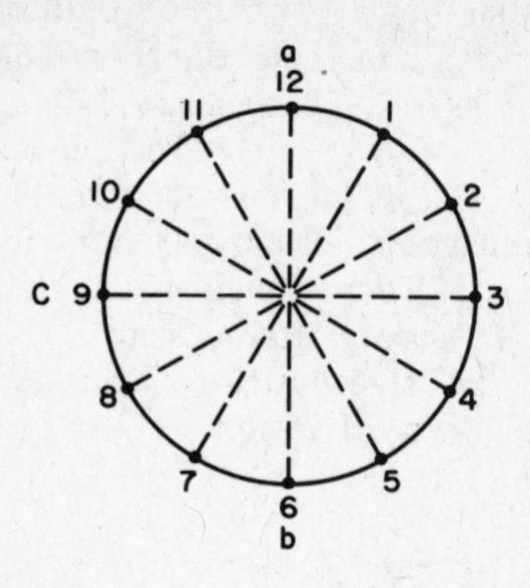

The men and women sat alternately. Since seats 12 and 6 were occupied by women, the men occupied seats 1, 3, 5, 7, or 11, while the women occupied seats 2, 4, 8, or, 10, for Mrs. Adams and Mrs. Burke were already seated at 6 and 12, and Mr. Clark was seated at 9. Since Mrs. Dexter was seated opposite Mrs. Foster and Mrs. Burke was seated opposite Mrs. Adams, Mrs. Clark must have been opposite Mrs. Egan. That is, a was opposite b, d was opposite f, and c was opposite e. But, by 4, Mrs. Egan was two places from Mrs. Adams, thus Mrs. Egan may have occupied seats 2 or 10. It follows that Mrs. Clark may have occupied seats 8 or 4, which are opposite Mrs. Egan. But Mrs. Clark cannot have occupied seat 8, for then she would have been next to her husband who was seated at 9. Therefore, she must have been seated at 4. But, if Mrs. Clark was seated at 4, Mrs. Egan must have been seated at 10, which is opposite seat 4. That is, c was in seat 4 and e was in seat 10. Moreover, by 3, Mr. Dexter was three places to the left of Mrs. Clark and, since Mrs. Clark was in seat 4, Mr. Dexter was in seat 7. Thus, so far, we have the result shown below.

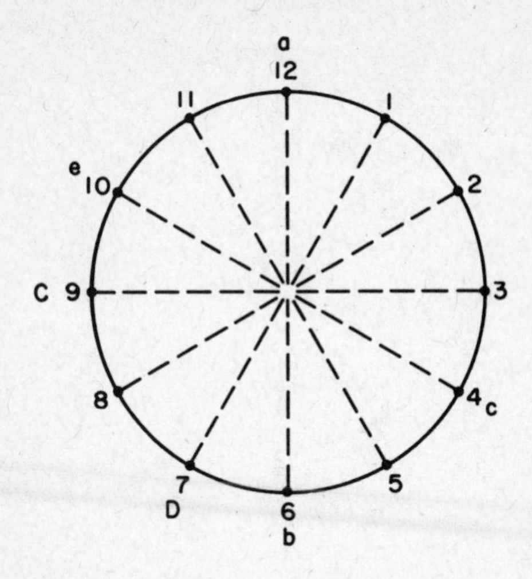

Seats 2 and 8 are now left for Mrs. Dexter and Mrs. Foster, but Mrs. Dexter could not have been in seat 8, for then she would have been next to her husband. Hence, Mrs. Dexter was in seat 2 and Mrs. Foster was in seat 8.

The husbands were separated from their wives by the same number of places. But Mrs. Dexter was five places away from Mr. Dexter. Thus, seat 11 must have been occupied by Mr. Clark or Mr. Burke. But Mr. Clark was in seat 9; hence, Mr. Burke must have been in seat 11. Seat 1 could have been occupied by Mr. Burke or Mr. Foster, but Mr. Burke was in seat 11; hence, Mr. Foster must have been in seat 1. Seat 3 may have been occupied by Mr. Foster or Mr. Egan, but Mr. Foster was in seat 1; hence, Mr. Egan was in seat 3. Finally, seat 5 could have been occupied by Mr. Egan or Mr. Adams, but Mr. Egan was in seat 3; hence, Mr.

Adams was in seat 5. The twelve people were seated as shown below.

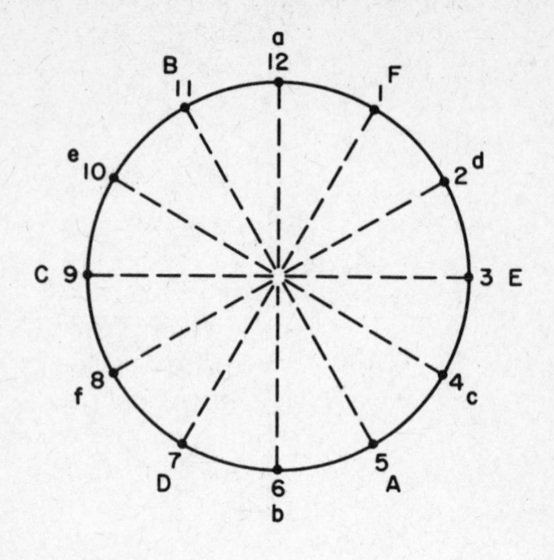

7. el camino del diablo

The men needed half a day to travel 5 miles from Sonoita, deposit $1\frac{1}{2}$ days' rations at the 5 mile point, and return to Sonoita. Then, they needed one full day to travel 20 miles: they journeyed to the 10 mile point, deposited $1\frac{1}{2}$ days' rations there, and returned to Sonoita, consuming $\frac{1}{2}$ day's rations at the 5 mile point, where they left the remaining 1 day's rations. Thus far, the men have spent one and a half days and established 1 day's rations at the 5 mile point and $1\frac{1}{2}$ days' rations at the 10 mile point.

Next, they traveled 20 miles in one day carrying 2 days' rations of which they consumed 1 day's rations and left 1 day's rations at the 20 mile point. On their return trip, which took one day, they consumed $\frac{1}{2}$ day's rations at both the 10 and 5 mile points, leaving ½ day's rations at the 5 mile point and 1 day's rations at the 10 mile point. So far, the men have traveled three and a half days and have deposits of ½ day's rations at the 5 mile point, and 1 day's rations each at the 10 and 20 mile points.

On their next trip, the men went as far as 30 miles in one and a half days. They consumed 1½ days' rations and deposited ½ day's rations at the 30 mile point. On their return trip, they took another one and a half days and consumed ½ day's rations at each of the 20, 10, and 5 mile points. Now, they have so far traveled a total of six and a half days and have deposits of ½ day's rations at each of the 10, 20, and 30 mile points, with nothing left at the 5 mile point. Since the same number of days are required to *double* the amount of the rations at the 10, 20, and 30 mile points, the men must then travel a total of thirteen days to establish deposits of 1 day's rations at each of these points.

The expedition then made its final 50 mile trip from Sonoita, carrying 2 days' rations. On the first day, they traveled to the 20 mile point and consumed ½ day's rations at both the 10 and 20 mile points. Thus, they left ½ day's rations at each of these points. On the second day, they traveled to the 40 mile point and they consumed ½ day's rations from the provisions left at the 30 mile point and ½ day's rations from the 2 days' rations which they were carrying. Thus, they left ½ day's rations at the 30 mile point and carried 1½ days' rations to the 40 mile point. On the third day, they consumed ½ day's rations of the 1½ days' rations carried at the 40 mile point, left ½ day's rations there, and proceeded to the 50 mile point with the remaining ½ day's rations. There, they consumed this ½ day's rations and returned to the 40 mile point, where they camped over night. On the fourth day, they consumed the $\frac{1}{2}$ day's rations which they had left at the 40 mile point and proceeded to the 20 mile point, again consuming the $\frac{1}{2}$ day's rations which they had left at the 30 mile point. On the fifth, and last, day, they returned to Sonoita, consuming the remaining $\frac{1}{2}$ day's rations which they had left at both the 20 and 10 mile points. Therefore, the men needed a total of $13 + 5 = 18$ days to complete their expeditions.

8. a hat trick

While the king was placing the hats on each man's head and the blindfolds were removed, Wise reasoned: the king must place more than one white hat on all of us, for if he placed a white hat on only one, or none of us, no one would raise their right hands when the blindfolds are removed, and one of the men wearing a black hat on seeing no white hats or only one white hat and no hands raised would conclude at once that he has a black hat. Moreover, the king must place at least three white hats, for if he places only two, the two men wearing them would not see more white than black hats and would not raise their right hands, but they would immediately realize that their hats are white. Furthermore, if the king places only three white hats and all the men raise their right hands, one of them wearing a white hat would conclude at once that he is wearing a white hat, since he would see a black hat and there can be no more than three white hats; in that case, not all will raise their right hands. If the king places three white hats and one black hat on our heads, the test would not be the same for each man; hence, it would not be a fair test. Thus, as soon as the blindfolds were removed and each of the men raised his right hand, Wise stood up and said, "Your Majesty, I am wearing a white hat."

9. a hat trick for the ladies

Pat reasoned as follows: If Carol and I had black caps, Ann would have known that her cap was green, because there are only two black caps. Since Ann did not know the color of her cap, then either Carol and I both have green caps or one of us has a green cap and the other a black one. If I had a black cap, Carol would have deduced that she had a green cap, for, otherwise, Ann would have known that she had a green cap. So the fact that Carol did not know the color of her cap made it clear to me that my cap was green.

10. the yard master's problem

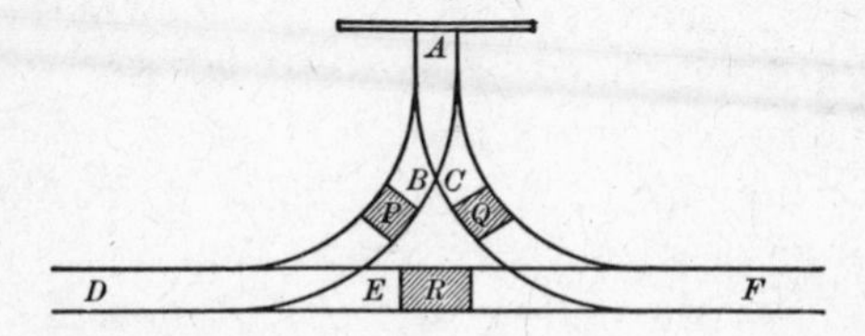

Engine R pushes car P into track A. Then engine R returns and pushes car Q to car P and couples cars P and Q. Next, engine R pulls cars P and Q out to track F, and then pushes them onto track E. Cars P and Q are then uncoupled and engine R returns car Q to track A, leaving car Q there. Now, engine R returns to car P and takes this car to track C, leaving car P there. Finally, engine R returns to tracks B and A, and pulls car Q onto track B.

11. the master detective

Let T and F denote *true* and *false*, respectively. Harry the Hawk is innocent, for if he were guilty, his statements (1) and (3) would both be false, contrary to the hypothesis that only one of each man's statements is false. Thus, a T is placed opposite Harry the Hawk in column 1 in the table. It follows that either Spanish Joe or Ricky the Rock is guilty. Assume that Ricky the Rock is guilty. Then Ricky's statement (1) is false and Spanish Joe's statement (1) is true. Thus, an F is placed opposite Ricky the Rock in column 1 and a T opposite Spanish Joe in the same column. It follows Spanish Joe's statement (3) is false; therefore, an F is placed opposite Spanish Joe in column 3. Now, since only one of each man's statements is false, then a T is placed opposite Spanish Joe in column 2 and T's opposite Ricky the Rock in columns 2 and 3. But Ricky's statement (2) says he has never seen Spanish Joe and Spanish Joe's statement (2) says Ricky is his pal. Hence, they cannot both be true, for they contradict each other. Therefore, the assumption that Ricky the Rock is guilty is false; hence, Ricky the Rock is innocent. Therefore, Spanish Joe must be the murderer. The reader may check this conclusion by assuming that Spanish Joe is guilty and showing that this assumption is consistent with the conditions and statements.

Statement	1	2	3
Harry the Hawk	T		
Spanish Joe	T	T	F
Ricky the Rock	F	T	T

12. d. tect gets his man

Let T and F denote *true* and *false*, respectively. Since each person made one correct statement, there are five correct statements. Now, there are only two different statements in statements (3) and one of these must be true; hence, the correct statement must be "blue eyes."

For, if brown eyes were true, since it occurs three times, it would leave only two more correct statements to answer the other three questions. However, this is impossible, for each question has been answered correctly at least once. Hence, "blue eyes" is true. Therefore, a T is placed in each of the rows in the table below headed Brown and O'Neil in column 3 and F's in all the other rows. Since each witness made only one correct statement, then F's are placed opposite Brown and O'Neil in all the other columns. There are only five correct statements and we have already used two. Hence, there remain three correct statements needed to answer the three remaining questions. Since each question has been answered correctly at least once, then only one statement is correct in each of the remaining questions. It follows that the statement, "dark blue," in statements (4) is false, for it occurs twice. Therefore, an F is placed opposite each of the rows headed Lash and Smith in column 4. It immediately follows that "black jacket" is correct, for at least one of the statements (4) is true. Therefore, a T is placed opposite Cohen in column 4 and F's in all the other columns. Now, from the table, "brown hair" is false; hence, an F is placed opposite Smith in column 2. It immediately follows that the statement, "blond hair," is correct; hence, a T is placed opposite Lash in column 2 and an F in column 1. Finally, from the table, "25" is correct so a T is placed opposite Smith in column 1. Thus, the guilty man was 25 years old, had blond hair and blue eyes, and wore a black jacket.

Statement	1	2	3	4
Brown	F	F	T	F
Cohen	F	F	F	T
O'Neil	F	F	T	F
Lash	F	T	F	F
Smith	T	F	F	F

13. murder in the back room

Let T and F denote true and false, respectively. If we assume that Lamb is guilty, then Lamb's statement (1) is false. Since by hypothesis only one of each man's statements is false, then statements (2) and (3) are true. Therefore, an F is placed in the table opposite Lamb in column 1 and T's in the same row in the other two columns. Since, under the assumption that Lamb is guilty, Lamb's statement (3) is true, then Clod's statement (3) must be false; hence, Clod's other two statements are true. Therefore, an F is placed opposite Clod in column 3 and T's in the same row in the other two columns. Now Dolt's statement (2) contradicts Lamb's statement (2); thus, Dolt's statement (2) is false. Moreover, Dolt's statement (3) contradicts Lamb's statement (3); hence, Dolt's statement (3) is false. This contradicts the hypothesis that only one of each man's statements is false. Thus, the assumption that Lamb is guilty is false, therefore he is innocent.

Statement	1	2	3
Lamb	F	T	T
Clod	T	T	F
Dolt		F	F

Now, assume that Clod is guilty. Then Dolt's statement (1) is false and, by hypothesis, his other two statements must be true. Therefore, an F is placed opposite Dolt in column 1 and T's in the same row in the other two columns. Since Lamb's statement (2) contradicts Dolt's statement (2), then from the table Lamb's statement (2) is false. Moreover, Lamb's statement (3) contradicts Dolt's statement (3); thus, Lamb's statement (3) is also false. This contradicts the hypothesis that only one of each man's statements is false. Hence, the assumption that Clod is guilty is also false; therefore, Clod is innocent.

Statement	1	2	3
Lamb		F	F
Clod			
Dolt	F	T	T

Finally, assume that Dolt is guilty, then his statement (2) is false and, by hypothesis, his other two statements must be true. Therefore, an F is placed opposite Dolt in column 2 and T's in the same row in the other two columns. Lamb's statement (3) contradicts Dolt's statement (3). Thus, Lamb's statement (3) is false and, by hypothesis, his other two statements are true. Therefore, an F is placed opposite Lamb in column 3 and T's in the same row in the other two columns. Clod's statement (3) contradicts Lamb's statement (3) and, since Lamb's statement (3) is false, then Clod's statement (3) is true. Therefore, a T is placed opposite Clod in column 3. Finally, Clod's statement (1) contradicts Dolt's statement (2) and, since Dolt's statement (2) is false, then Clod's statement (1) must be true. Therefore, a T is placed opposite Clod in column 2. It follows that Clod's statement (2) is false and, since this does not contradict any of the other statements, an F is placed opposite Clod in column 2. Thus, the assumption that Dolt is guilty is consistent with our hypothesis and with all the statements.

Statement	1	2	3
Lamb	T	T	F
Clod	T	F	T
Dolt	T	F	T

14. d. tect cried murder

Let T and F denote *true* and *false,* respectively. First, assume that Knave is guilty. Since both statements made by the criminal are false, F's are placed opposite Knave in both columns of the table. It immediately follows from the fact that Knave's first statement is false, that Divine is the criminal. But, this is impossible, for we have assumed that Knave is the criminal. Hence, our assumption is wrong; therefore, Knave is innocent.

Statement	1	2
Knave	F	F
Cell		
Divine		

Now, assume that Divine is guilty; then, by hypothesis, both of his statements are false. Therefore, F's are placed opposite Divine in both columns. It immediately follows that both of the statements made by Knave are also false. Therefore, F's are placed opposite Knave in both columns. However, this contradicts the hypothesis that the criminal is the only one whose statements are both false. Hence, the assumption that Divine is guilty is also incorrect; therefore Divine is innocent. Thus, Cell must be guilty.

Statement	1	2
Knave	F	F
Cell		
Divine	F	F

This conclusion can easily be seen by assuming that Cell is guilty. Then, by hypothesis, both of his statements are false. Therefore, F's are placed opposite Cell in both columns. It immediately follows that both of the statements made by Knave are true. Therefore, T's are placed opposite Knave in both columns. Moreover, the first statement made by Divine is true. Therefore, a T is placed opposite Divine in the first column. Finally, the second statement made by Divine is false. Therefore, an F is placed opposite Divine in the second column. This combination agrees with the hypothesis. Hence, Cell is guilty. It follows that Knave is the clergyman, Cell is the electrician, and Divine is the ex-convict.

Statement	1	2
Knave	T	T
Cell	F	F
Divine	T	F

15. who's who?

The solution of problems containing what seems to be a series of unrelated statements is made easier by making a table of all possibilities or an array of possible results and then examining the consistency of the statements. Thus, in this problem, you can form the following table.

METHOD I

Profession	Physicist	Philosopher	Pianist
Clark			
Morris			
Howe			

After examining the statements, if you conclude that Morris cannot be the pianist, place an X opposite his name in the column headed Pianist. If you deduce that Clark has to be the philosopher, place an O opposite his name in the column headed Philosopher.

METHOD II

You can also consider all the possible combinations and, by examining the statements, eliminate the ones that are not consistent, until only one possibility remains. Thus, in this problem, if H, M, and C denote Howe, Morris, and Clark, respectively, the possible combinations are:

Possibility No.	1	2	3	4	5	6
Physicist	C	C	M	M	H	H
Pianist	M	H	C	H	M	C
Philosopher	H	M	H	C	C	M

If after examining the statements, you conclude that Morris cannot be the pianist, this immediately eliminates possibilities 1 and 5.

Solution by method I. By 3, both the philosopher and the physicist know the pianist and, by 2, Clark has never heard of Morris so they do not know each other. Hence, Howe must be the pianist. Thus, an O is placed in the column headed Pianist opposite Howe. Since Howe is the pianist, he cannot be the physicist or the philosopher, so X's are placed opposite Howe in the columns headed Physicist and Philosopher. Moreover, neither Clark nor Morris is the pianist; hence, X's are placed opposite these names in the column headed Pianist. By 1, Morris makes more money than Clark

	Physicist	Philosopher	Pianist
Clark			X
Morris			X
Howe	X	X	O

and, by 4, the pianist makes more money than the philosopher. Therefore, Morris is the Physicist. Thus, an O is placed opposite Morris in the column headed Physicist, and an X in the same column opposite Clark. Since Morris is the physicist, he cannot be the philosopher, so an X is placed opposite Morris in the column headed Philosopher. It immediately follows that Clark is the philosopher. Therefore, Howe is the pianist, Morris is the physicist, and Clark is the philosopher.

	Physicist	Philosopher	Pianist
Clark	X	O	X
Morris	O	X	X
Howe	X	X	O

Solution by method II. If H, M, and C denote Howe, Morris, and Clark, respectively, all possible combination are shown in this table.

Possibility No.	1	2	3	4	5	6
Physicist	C	C	M	M	H	H
Pianist	M	H	C	H	M	C
Philosopher	H	M	H	C	C	M

By 3, both the philosopher and the physicist know the pianist and, by 2, Clark has never heard of Morris, so they do not know each other; hence, Howe must be the pianist. This conclusion eliminates possibilities that do not have Howe as the pianist; that is, it eliminates possibilities 1, 3, 5, and 6. By 1, Morris makes more money than Clark and, by 4, the physicist makes more money than the philosopher; hence, Morris is the physicist. This eliminates possibility 2. Therefore, the only combination that is consistent with the given statements is possibility 4; therefore, Morris is the physicist, Howe is the pianist, and Clark is the philosopher.

16. the bridge game

Solution by method I. (See No. 15.) By 1, Bill plays with Grace as a partner and Jack with Becky. Hence, Grace is not Bill's wife and Jack is not Becky's husband; therefore, an X is placed opposite Bill in the column headed Grace and an X opposite Jack in the column headed Becky. By 2, Joan plays with Jack as a partner, so that Joan is not Jack's wife; therefore, an X is placed opposite Jack in the column headed Joan. It immediately follows, from the table, that Grace is Jack's wife; therefore, an O is placed opposite Jack in the column headed Grace and an X in the same column opposite Sam, for then Grace cannot be Sam's wife.

Now Bill must be married either to Becky or Joan. If we assume that Bill is married to Becky, then Sam must be married to Joan. But Bill plays with Sam's wife, in this case Joan, as a partner and, by 2, Joan plays with Jack as a partner, which is impossible. Thus, Bill must be married to Joan; hence an O is placed opposite Bill in the column headed Joan an X in the same row in the column headed Becky and another X opposite Sam in the column headed Joan. It immediately follows that Becky is Sam's wife. Therefore, an O is placed opposite Sam in the column headed Becky. Thus, Bill is married to Joan, Sam is married to Becky, and Jack is married to Grace.

Players	Grace	Joan	Becky
Bill	X	O	X
Sam	X	X	O
Jack	O	X	X

Solution by method II. (See No. 15.) If G, J, and B denote Grace, Joan, and Becky, respectively, the possible combinations are shown in the table.

Possibility No.	1	2	3	4	5	6
Bill	G	G	J	J	B	B
Sam	J	B	G	B	G	J
Jack	B	J	B	G	J	G

By 1, Bill plays with Grace as a partner. Hence, Grace is not Bill's wife. This conclusion eliminates possibilities 1 and 2. Again, by 1, Jack plays with Becky as a partner; thus, Becky is not Jack's wife. This eliminates possibility 3. By 2, Joan plays with Jack as a partner; hence, Joan is not Jack's wife. This eliminates possibility 5. Now, from the table Bill must be married either to Joan or Becky. If we assume Bill is married to Becky, then Sam must be married to Joan. But, by 3, Bill plays with Sam's wife (in this case Joan) as a partner and, by 2, Joan plays with Jack as a partner, which is impossible; therefore, Becky is not Bill's wife. This eliminates possibility 6. Hence, the only remaining possibility that is consistent with the given statements is possibility 4. Therefore, Bill, Sam, and Jack are the respective husbands of Joan, Becky, and Grace.

17. who verifies the accounts?

Solution by method I. (See No. 15.) By 2, Baker is married and, by 1, the treasurer is single; hence, Baker cannot be the treasurer. Therefore, an X is placed op-Baker in the column in the table headed Treasurer. By 1, the treasurer earns the least and, by 3, Ennis earns more money than the manager; hence, Ennis cannot be the treasurer. Therefore, an X is placed opposite Ennis in the column headed Treasurer. It immediately follows from the table that Cronin is the treasurer; therefore, an O is placed opposite Cronin in the column headed Treasurer. Since Cronin is the treasurer, he cannot be the controller nor the manager, so X's are placed opposite Cronin in the columns headed Manager and Controller. Now, by 3, Ennis earns more money than the manager, so he cannot be the manager; thus, Ennis is the controller. Therefore Cronin, Baker, and Ennis are the treasurer, manager, and controller, respectively.

Officer	Controller	Manager	Treasurer
Cronin	X	X	O
Baker	X	O	X
Ennis	O	X	X

Solution by method II. (See No. 15.) If C, B, and E denote Cronin, Baker, and Ennis, respectively, the possible combinations are given in this table.

Possibility No.	1	2	3	4	5	6
Controller	C	C	B	B	E	E
Manager	B	E	C	E	B	C
Treasurer	E	B	E	C	C	B

By 2, Baker is married and, by 1, the treasurer is single; hence, Baker cannot be the treasurer. This eliminates possibilities 2 and 6. By 1, the treasurer earns the least and, by 3, Ennis earns more money than the manager; hence, Ennis cannot be the treasurer. Thus, possibilities 1 and 3 are eliminated. By 3, Ennis earns more money than the manager, so Ennis cannot be the manager and possibility 4 is eliminated. Therefore, 5 is the only possibility consistent with the statements, and so Ennis, Baker, and Cronin are the controller, manager, and treasurer, respectively.

18. untie this nuptial knot

Solution by method I. (See No. 15.) By 3, Pete does not play golf and, by 1, Jill's and Flo's husbands play golf; hence, Pete is not married to Jill or to Flo. Therefore, an X is placed opposite Pete in the columns headed Jill and Flo. It immediately follows that Pete is married to Cora; therefore, an O is placed opposite Pete in the column headed Cora. By 1, Dan's wife plays with Jill's husband and, by 2, husband and wife do not play as partners; hence, Dan is not Jill's husband. Therefore, an X is placed opposite Dan in the column headed Jill. It immediately follows from the table that Sid is married to Jill. Thus, place an O opposite Sid in the column headed Jill and, as a consequence, an X opposite Sid in the column headed Flo. Hence, Dan, Pete, and Sid are the respective husbands of Flo, Cora, and Jill.

Players	Cora	Jill	Flo
Dan	X	X	O
Pete	O	X	X
Sid	X	O	X

Solution by method II. (See No. 15.) If C, J, and F denote Cora, Jill, and Flo, respectively, the possible combinations are given in this table.

Possibility No.	1	2	3	4	5	6
Pete	C	C	J	J	F	F
Dan	J	F	C	F	C	J
Sid	F	J	F	C	J	C

By 3, Pete is not a golf player, and, by 1, both the husbands of Jill and Flo play golf. Hence, Pete is neither married to Jill nor to Flo, so he must be married to Cora. This conclusion immediately eliminates possibilities 3, 4, 5, and 6. By 1, Dan's wife plays golf with Jill's husband and, by 2, husband and wife do not play as partners; hence, Dan is not Jill's husband. This fact eliminates possibility 1. Therefore, 2 is the only possibility consistent with the given statements. Hence, Pete, Dan, and Sid are the respective husbands of Cora, Flo, and Jill.

19. the train puzzle

Solution by method I. (See No. 15.) By 4, Smith cannot be the fireman. Therefore, an X is placed opposite his name in the column headed Fireman. By 3, the brakeman lives halfway between Chicago and Detroit. Thus, his next door neighbor, one of the passengers, cannot live in either Chicago or Detroit; hence, by 1, he cannot be Mr. Robinson. Moreover, by 5, his next door neighbor earns exactly three times as much yearly as the brakeman, but, by 2, the sum earned by Mr. Jones is

Profession	Engineer	Fireman	Brakeman
Jones	X	X	O
Robinson			X
Smith		X	X

exactly $20,000 and this sum is not exactly divisible by three. Thus, Mr. Jones cannot be the brakeman's next door neighbor; therefore, it must be Mr. Smith. It follows that Mr. Smith cannot live in Chicago or Detroit. Now, by 1, Mr. Robinson lives in Detroit; hence, Mr. Jones must live in Chicago; hence, by 6, Jones is the brakeman. Therefore, an O is placed opposite Jones in the column headed Brakeman. Since Jones is the brakeman, he can neither be the engineer nor the fireman. Therefore, X's are placed opposite Jones in the columns headed Engineer and Fireman. Moreover, because Jones is the brakeman, neither Robinson nor Smith can be the brakeman; hence X's are placed in the column headed Brakeman opposite the names Robinson and Smith. The table, above, now makes it evident that Robinson is the fireman. Therefore, an O is placed opposite Robinson in the column headed Fireman and an X opposite Robinson in the column headed Engineer and, as a consequence, an O in the same column opposite Smith. Thus, Jones is the brakeman, Robinson is the fireman, and Smith is the engineer.

Profession	Engineer	Fireman	Brakeman
Jones	X	X	O
Robinson	X	O	X
Smith	O	X	X

Solution by method II. (See No. 15.) If J, R, and S denote Jones, Robinson, and Smith, respectively, the possible combinations are given in this table.

Possibility No.	1	2	3	4	5	6
Engineer	J	J	R	R	S	S
Fireman	R	S	J	S	J	R
Brakeman	S	R	S	J	R	J

By 4, Smith cannot be the fireman, so we eliminate possibilities 2 and 4. Reasoning as in method I, we conclude that Jones is the brakeman. This conclusion eliminates possibilities 1, 3, and 5, that do not have Jones as the brakeman. Therefore, 6 is the only possibility consistent with the given statements. Thus, Smith is the engineer, Robinson is the fireman, and Jones is the brakeman.

20. it's a wise son who knows his own father

Solution by method I. (See No. 15.) By 1, Ed's child is a girl and, by 2, Al's child is also a girl; hence, Fred's child must be a boy and Tony is Fred's son. Therefore, an O is placed in the column headed Tony opposite Fred, and X's in the same column opposite Al and Ed. By 2, Al's daughter is not Jane. Therefore, an X is placed in the column headed Jane opposite Al. It immediately follows from the table that Mary is Al's daughter; therefore, an O is placed in the column headed Mary opposite Al and X's in the same column opposite Fred and Ed. Again, from the table Jane must be Ed's daughter; therefore, an O is placed in the column headed Jane opposite Ed and an X in the same column opposite Fred.

Now, Grace is not married to Fred, for Fred has a son, and, by 1, Grace has a daughter; therefore, an X is placed in the column headed Grace opposite Fred. By 3, Fred's wife is not Rose; therefore, an X is placed in the column headed Rose opposite Fred. It immediately follows from the table that Fred is married to Alice; therefore, an O is placed in the column headed Alice opposite Fred and X's in the same column opposite Al and Ed. By 2, Grace is not married to Ed, for each couple has only one child each; therefore, an X is placed opposite Ed in the column headed Grace. It immediately follows from the table that Al is married to Grace; therefore an O is placed in the column headed Grace opposite Al and an X in the same row in the column headed Rose. Hence, Ed is married to Rose; therefore, an O is placed in the column headed Rose opposite Ed. Thus, the families are: Al, Grace, and Mary; Fred, Alice, and Tony; and Ed, Rose, and Jane.

	Alice	Grace	Rose	Mary	Tony	Jane
Al	X	O	X	O	X	X
Fred	O	X	X	X	O	X
Ed	X	X	O	X	X	O

Solution by method II. (See No. 15.) If A, G, R, M, T, and J denote Alice, Grace, Rose, Mary, Tony, and Jane, respectively, the possible combinations are given in this table.

	Wives						Children					
Possibility No.	1	2	3	4	5	6	7	8	9	10	11	12
Al	A	A	G	G	R	R	M	M	T	T	J	J
Fred	G	R	R	A	A	G	T	J	J	M	M	T
Ed	R	G	A	R	G	A	J	T	M	J	T	M

By 1, Ed's child is a girl. This statement eliminates possibilities 8 and 11 which show that Tony is Ed's child. By 2, Al's child is also a girl; thus, possibilities 9 and 10 are eliminated. By 2, Al's daughter is not Jane, eliminating possibility 12. This leaves 7 as the only possibility for the children.

Grace is not married to Fred, for Fred has a son and, by 1, Grace has a daughter. This statement eliminates possibilities 1 and 6. By 3, Fred's wife is not Rose, eliminating possibilities 2 and 3. By 2, Grace is not married to Ed, for each couple has one child only. Thus, possibility 5 is eliminated, leaving 4 as the only possibility for the wives. Therefore, the possibilities are 4 and 7. Hence, the families are: Al, Grace, and Mary; Fred, Alice, and Tony; and Ed, Rose, and Jane.

21. the professionals

Solution by method I. (See No. 15.) By 2, Ann has never heard of the vanishing point in perspective drawing; hence, she can be neither the architect nor the draftsman. Therefore, an X is placed opposite Ann in each of the columns headed Draftsman and Architect. By 1, the doctor is married and, by 3, Bob is a bachelor; thus, Bob cannot be the doctor. Therefore, an X is placed opposite Bob in the column headed Doctor. By 1, the architect is the doctor's wife; hence, it is either Maggie or Ann. But, by 2, Ann cannot be the architect; hence, Maggie is the architect. Therefore, an O is placed opposite Maggie in the column headed Architect and X's in the same row in the other columns. Moreover, since Maggie is the architect, neither Henry nor Bob can be the architect. Therefore, X's are placed in the column headed Architect opposite Henry and Bob. By 1, the doctor is married to Maggie, the architect, so he is a man; thus, Ann cannot be the doctor. Therefore, an X is placed opposite Ann in the column headed Doctor. It immediately follows from the table that Henry is the doctor; therefore, an O is placed in the column headed Doctor opposite Henry and X's in the same row in the other columns. Again, from the table, it follows that Bob is the draftsman; therefore, an O is placed opposite Bob in the column headed Draftsman and an X in the same row in the column headed Lawyer; hence Ann must be the lawyer. Thus, Henry is the doctor, Bob is the draftsman, Maggie is the architect, and Ann is the lawyer.

	Doctor	Draftsman	Architect	Lawyer
Henry	O	X	X	X
Bob	X	O	X	X
Maggie	X	X	O	X
Ann	X	X	X	O

Solution by method II. (See No. 15.) Since there are 24 possible combinations, this method becomes rather awkward. However, for the sake of completeness, it will be given. If H, B, M, and A denote Henry, Bob, Maggie, and Ann, respectively, the possible combinations are given in this table.

Possibility No.	1	2	3	4	5	6	7	8	9	10	11	12
Doctor	H	H	H	H	H	H	B	B	B	B	B	B
Draftsman	B	B	M	M	A	A	H	H	M	M	A	A
Architect	M	A	B	A	B	M	M	A	H	A	H	M
Lawyer	A	M	A	B	M	B	A	M	A	H	M	H

Possibility No.	13	14	15	16	17	18	19	20	21	22	23	24
Doctor	M	M	M	M	M	M	A	A	A	A	A	A
Draftsman	H	H	B	B	A	A	H	H	B	B	M	M
Architect	B	A	H	A	H	B	B	M	H	M	H	B
Lawyer	A	B	A	H	B	H	M	B	M	H	B	H

By 2, Ann has never heard of the vanishing point in perspective drawing; hence, she can be neither the architect nor the draftsman. Therefore, cross out possibilities 2, 4, 5, 6, 8, 10, 11, 12, 14, 16, 17, and 18. By 1, the doctor is married and, by 3, Bob is a bachelor; hence, Bob cannot be the doctor. This deduction eliminates possibilities 7 and 9. By 1, the architect is the wife of the doctor and, by 2, Ann cannot be the architect. Therefore, Maggie is the architect, eliminating possibilities 3, 20, and 22. Moreover, since Maggie is the architect, neither Henry nor Bob can be the architect. Hence, possibilities 13, 15, 19, 21, 23, and 24 are eliminated. This leaves 1 as the only possibility consistent with the statements. Therefore, Henry is the doctor, Bob is the draftsman, Maggie is the architect, and Ann is the lawyer.

22. the executive officers

Solution by method I. (See No. 15.) By 1, Blair is not the president; therefore, an X is placed opposite Blair in the column headed President. By 2, Koch is also not the president; therefore, an X is placed opposite Koch in the column headed President. By 5, Lewis served in the navy and, by 1, the president served in the army; hence, Lewis is not the president. Therefore, an X is placed opposite Lewis in the column headed President. It immediately follows from the table that Warren is the president. Therefore, an O is placed opposite Warren in the column headed President and X's in the same row in all other columns.

By 3, Lewis is not the secretary; therefore, an X is placed opposite Lewis in the column headed Secretary. By 6, Koch does not play golf or cards, but, by 3, the secretary plays golf, and, by 4, the treasurer plays cards; hence, Koch is neither the secretary nor the treasurer. Therefore, X's are placed opposite Koch in the columns headed Secretary and Treasurer. It follows from the table that Blair is the secretary; therefore, an O is placed opposite Blair in the column headed

Secretary and X's in the same row in all other columns. Again, from the table Lewis must be the treasurer; therefore, an O is placed opposite Lewis in the column headed Treasurer and X's in the same row in all other columns. Hence, Koch is the vice-president; therefore, an O is placed opposite Koch in the column headed Vice-President. Therefore, the executive officers are: Warren, president; Koch, vice-president; Lewis, treasurer; and Blair, secretary.

	President	Vice-President	Treasurer	Secretary
Blair	X	X	X	O
Koch	X	O	X	X
Lewis	X	X	O	X
Warren	O	X	X	X

Solution by method II. (See No. 15.) If B, K, L, and W denote Blair, Koch, Lewis, and Warren, respectively, the possible combinations are given in this table.

Possibility No.	1	2	3	4	5	6	7	8	9	10	11	12
President	B	B	B	B	B	B	K	K	K	K	K	K
Vice-President	K	K	L	L	W	W	B	B	L	L	W	W
Treasurer	L	W	K	W	K	L	L	W	B	W	B	L
Secretary	W	L	W	K	L	K	W	L	W	B	L	B

Possibility No.	13	14	15	16	17	18	19	20	21	22	23	24
President	L	L	L	L	L	L	W	W	W	W	W	W
Vice-President	W	W	K	K	B	B	B	B	K	K	L	L
Treasurer	K	B	B	W	K	W	K	L	B	L	B	K
Secretary	B	K	W	B	W	K	L	K	L	B	K	B

By 1, Blair is not the president. This statement eliminates possibilities 1, 2, 3, 4, 5, and 6. By 2, Koch is not the president; this eliminates possibilities 7, 8, 9, 10, 11, and 12. By 5, Lewis served in the navy, and, by 1, the president served in the army; hence, Lewis is not the president. This deduction eliminates possibilities 13, 14, 15, 16, 17, and 18. By 3, Lewis is not the secretary; this eliminates possibilities 19 and 21. By 6, Koch does not play golf or cards, but, by 3, the secretary plays golf and, by 4, the treasurer plays cards; hence, Koch is neither the secretary nor the treasurer. Thus, possibilities 20, 23, and 24 are eliminated. This leaves 22 as the only possibility consistent with the statements. Therefore, the executive officers are: Warren, president; Koch, vice-president; Lewis, treasurer; and Blair, secretary.

23. whose little girl are you?

Solution by method I. (See No. 15.) No father and daughter took part together in the same match; hence, by 1, Mary and Joan are not the daughters of Robert or Edward. Therefore, X's are placed in the columns headed Joan and Mary opposite Robert and Edward. Similarly, by 3, Pat and Mary are not the daughters of William or Carl; therefore, X's are placed in the columns headed Pat and Mary opposite William and Carl. By 5, Joan is also not the daughter of William or Carl; therefore, X's are placed in the column headed Joan opposite William and Carl. By 8, Joan is not the daughter of Thomas; therefore, an X is placed in the column headed Joan opposite Thomas. It follows from the table that Joan is Arthur's daughter; therefore, an O is placed opposite Arthur in the column headed Joan, and X's in the same row in all other columns. Then, from the table Mary must be the daughter of Thomas; therefore, an O is placed opposite Thomas in the column headed Mary and X's in the same row in all other columns. By 10, Pat and Nancy are not Edward's daughters; therefore an X is placed opposite Edward in the columns headed Pat and Nancy. Then, from the table Robert is Pat's father; therefore, an O is placed opposite Robert in the column headed Pat and X's in the same row in all other columns. By 12, William is not the father of Helen or Nancy; therefore, an X is placed opposite William in the columns headed Helen and Nancy. It follows from the table that Carl is Nancy's father; therefore, an O is placed opposite Carl in the column headed Nancy and X's in the same row in all other columns. Now, from the table Edward is Helen's father; therefore, place an O opposite Edward in the column headed Helen and X's in the same row in all other columns. Finally, the table shows that William is Dorothy's father. Thus, the father-daughter relationship is: William and Dorothy, Robert and Pat, Thomas and Mary, Edward and Helen, Arthur and Joan, and Carl and Nancy.

	Pat	Joan	Dorothy	Nancy	Mary	Helen
William	X	X	O	X	X	X
Robert	O	X	X	X	X	X
Thomas	X	X	X	X	O	X
Edward	X	X	X	X	X	O
Arthur	X	O	X	X	X	X
Carl	X	X	X	O	X	X

By 2, Dorothy, Pat, Helen, and Thomas' daughter, Mary, are not from Pennsylvania; therefore, X's are placed in the column headed Penn. in the table opposite Dorothy, Pat, Helen, and Mary. By 4, Mary, Dorothy, Pat, and Carl's daughter, Nancy, are not from Connecticut; therefore, X's are placed in the column headed Conn. opposite Mary, Dorothy, Pat, and Nancy.

By 7, Dorothy is not from Virginia; therefore, an X is placed in the column headed Va. opposite Dorothy. By 9, Robert's daughter, Pat, and William's daughter, Dorothy, are not from New Jersey; therefore, X's are placed in the column headed N.J. opposite Pat and Dorothy. By 13, Nancy is not from Pennsylvania or New Jersey; therefore, X's are placed opposite Nancy in the columns headed Penn. and N.J. Then, from the table Joan is from Pennsylvania; therefore, an O is placed opposite Joan in the column headed Penn., and X's in the same row in all other columns. It follows from the table that Helen is from Connecticut; therefore, an O is placed opposite Helen in the column headed Conn. and X's in the same row in all other columns. Again, from the table Mary is from New Jersey; therefore, an O is placed opposite Mary in the column headed N.J. and X's in the same row in all other columns. By 15, neither Dorothy nor Nancy are from Massachusetts; therefore, X's are placed in the column headed Mass. opposite Dorothy and Nancy. It immediately follows from the table that Pat is from Massachusetts; therefore, an O is placed opposite Pat in the column headed Mass. and X's in the same row in all other columns. Then, from the table Nancy must be from Virginia; therefore, place an O opposite Nancy in the column headed Va. and X's in the same row in all other columns. Finally, the table shows that Dorothy is from New York. Therefore, the father-daughter relationship and their respective home states are: William, Dorothy, New York; Robert, Pat, Massachusetts; Thomas, Mary, New Jersey; Edward, Helen, Connecticut; Arthur, Joan, Pennsylvania; and Carl, Nancy, Virginia.

The solution by method II would be clumsy and rather long for there are 216 possible different combinations, therefore it has been omitted.

	N.J.	Va.	Conn.	Penn.	Mass.	N.Y.
Pat	X	X	X	X	O	X
Joan	X	X	X	O	X	X
Dorothy	X	X	X	X	X	O
Nancy	X	O	X	X	X	X
Mary	O	X	X	X	X	X
Helen	X	X	O	X	X	X

24. the traveling salesmen

(See No. 15.) Let C, T, G, H, B, and J denote carpets, toys, glassware, hardware, boats, and jewelry, respectively. Then, by 1, Stark does not sell hardware; therefore, an X is placed opposite Stark in column H in the table. By 3, Hart does not sell hardware or jewelry; therefore, an X is placed opposite Hart in columns H and J. Since the salesmen sit three on each side, then one side consists of the hardware salesman, Hart, and the jewelry salesman. Call this row 1.

Row 1		
H	Hart	J

By 4, the hardware salesman sits opposite the toy salesman; thus, the toy salesman is in row 2. Hence, Hart does not sell toys; therefore, an X is placed opposite Hart in column T. By 5, Carter does not sell carpets; therefore, an X is placed opposite Carter in column C. By 6, Hart is not the glassware salesman; therefore, an X is placed opposite Hart in column G. By 7, Waters does not sell carpets; therefore, an X is placed opposite Waters in column C. By 8, Stark does not sell toys; therefore, an X is placed opposite Stark in column T. By 9, Carter does not sell glassware; therefore, an X is placed opposite Carter in column G.

Since Hart is not the glassware salesman, the glassware salesman must be in row 2 with the toy salesman. Again, by 9, Carter sits opposite the glassware salesman; hence, Carter sits in row 1. It follows Carter is either the hardware salesman or the jewelry salesman, but the toy salesman sits opposite the hardware sales man and, by 9, Carter sits opposite the glassware salesman. Hence, Carter cannot be the hardware salesman; thus, he must be the jewelry salesman. Therefore, an O is placed opposite Carter in column J and X's in the same row in all other columns; also, X's are placed in column J in all other rows.

Stark is not the hardware salesman, and, by 8, he sits at the end of a row. Now he cannot be the jewelry salesman, for we have just deduced that Carter is the jewelry salesman; hence, Stark must be in row 2. But, by 8, Stark is not the toy salesman, and, since, by 4, the toy salesman sits opposite the hardware salesman, Stark must sit opposite the jewelry salesman, Carter. Now, by 9, the glassware salesman sits opposite Carter; hence, Stark is the glassware salesman. Therefore, an O is placed opposite Stark in column G and X's in the same row in all other columns; also, X's in column G in all other rows.

Now, from the table Hart must be either the carpet or boat salesman. Assume that Hart is the carpet salesman; therefore an O is placed opposite Hart in column C and X's in the same row in all other columns. Also, X's are placed in column C in all other rows. Furthermore, since, by 7, Waters sits next to the carpet salesman, he must be the hardware salesman, because we know that Carter, the jewelry salesman, sits next to Hart. Therefore, an O is placed in column H opposite Waters and X's in the same row in all other columns; also, X's are placed in column H in all other rows.

	C	T	G	H	B	J
Stark	X	X	O	X	X	X
Hart	O	X	X	X	X	X
Duffy	X		X	X		X
Carter	X	X	X	X	X	O
Waters	X	X	X	O	X	X
Ring	X		X	X		X

Now, from the table Duffy must be either the toy salesman or the boat salesman. If Duffy is the toy salesman then, by 2, Duffy would have to be reading the hardware catalog. Hence, Duffy cannot be the toy salesman; thus, he must be the boat salesman. However, if he were the boat salesman, he cannot be sitting opposite either the hardware salesman or Carter, the jewelry salesman, for we know Stark sits opposite Carter. Then, Duffy must be sitting opposite Hart and, by 2, he must be reading the carpet catalog. However, by 5, Carter is reading the carpet catalog; hence, Duffy cannot be either the toy or the boat salesman. Thus, Waters cannot be the hardware salesman and Hart cannot be the carpet salesman. Therefore, all the conclusions arrived at from the assumption that Hart is the carpet salesman are incorrect. Hence, erase all these conclusions.

Now, since Hart cannot be the carpet salesman, he must be the boat salesman; therefore, an O is placed opposite Hart in column B and X's in the same row in all other columns; also, X's are placed in column B in all other rows. From the table, Waters must either be the toy or the hardware salesman. Now, Waters cannot be the hardware salesman for the hardware salesman sits next to Hart who sells boats and, by 7, Waters sits next to the carpet salesman. Hence, Waters must be the toy salesman; therefore, an O is placed opposite Waters in column T and X's in the same row in all other columns; also, X's are placed in column T in all other rows. By 10, Ring is not interested in carpets; therefore, an X is placed opposite Ring in column C. It immediately follows from the table that Ring sells hardware; therefore, an O is placed opposite Ring in column H and X's in the same column in all the other rows. Finally, from the table Duffy must sell carpets; therefore, place an O opposite Duffy in column C. Thus, Stark sells glassware, Hart sells boats, Duffy sells carpets, Carter sells jewelry, Waters sells toys, and Ring sells hardware.

	C	T	G	H	B	J
Stark	X	X	O	X	X	X
Hart	X	X	X	X	O	X
Duffy	O	X	X	X	X	X
Carter	X	X	X	X	X	O
Waters	X	O	X	X	X	X
Ring	X	X	X	O	X	X

25. the performing arts

Solution by method I. (See No. 15.) By 1, neither Marinelli nor Hall is the pianist; therefore, X's are placed in the column headed Pianist opposite Marinelli and Hall. By 2, Pastore cannot be the writer or the painter; therefore, X's are placed opposite Pastore in the columns headed Writer and Painter. By 3, the writer is neither Reno nor Marinelli; therefore, X's are placed in the column headed Writer opposite Reno and Marinelli. It follows from the table that Hall is the writer. Therefore, an O is placed opposite Hall in the column headed Writer and X's in the same row in all other columns. By 2, Hall sat for the painter, while, by 4, Marinelli does not know Hall; hence, Marinelli cannot be the painter. Therefore, an X is placed opposite Marinelli in the column headed Painter. Then, from the table Reno is the painter; therefore, an O is placed opposite Reno in the column headed Painter and X's in the same row in all other columns. It immediately follows from the table that Pastore is the pianist. Therefore, an O is placed opposite Pastore in the column headed Pianist and an X in the same row in the column headed Dancer. It immediately follows from the table that Marinelli is the dancer. Thus, Marinelli is the dancer, Reno is the painter, Hall is the writer, and Pastore is the pianist.

	Dancer	Painter	Writer	Pianist
Hall	X	X	O	X
Reno	X	O	X	X
Marinelli	O	X	X	X
Pastore	X	X	X	O

Solution by method II. (See No. 15.) Let H, R, M, and P denote Hall, Reno, Marinelli and Pastore, respectively. Then, the different possibilities are given in the table.

Possibility No.	1	2	3	4	5	6	7	8	9	10	11	12
Dancer	H	H	H	H	H	H	R	R	R	R	R	R
Painter	R	R	M	M	P	P	H	H	M	M	P	P
Writer	M	P	R	P	R	M	M	P	H	P	H	M
Pianist	P	M	P	R	M	R	P	M	P	H	M	H

Possibility No.	13	14	15	16	17	18	19	20	21	22	23	24
Dancer	M	M	M	M	M	M	P	P	P	P	P	P
Painter	P	P	H	H	R	R	H	H	R	R	M	M
Writer	H	R	P	R	P	H	R	M	H	M	H	R
Pianist	R	H	R	P	H	P	M	R	M	H	R	H

By 1, neither Marinelli nor Hall is the pianist; this eliminates possibilities 2, 5, 8, 10, 11, 12, 14, 17, 19, 21, 22, and 24. By 2, Pastore cannot be the writer or the painter; this eliminates possibilities 4, 6, 13, 15, and 17. By 3, the writer is neither Reno nor Marinelli; this eliminates possibilities 1, 3, 7, 16, and 20 and leaves possibilities 9, 18, and 23. In each of these possibilities Hall is the writer. Hence, we now know that Hall is the writer. By 2, the writer, Hall, sat for the painter while, by 4, Marinelli has never heard of Hall; thus, Marinelli is not the painter. This eliminates possibilities 9 and 23; hence, the only possibility is 18, which is consistent with the statements. Therefore, Marinelli is the dancer, Reno is the painter, Hall is the writer, and Pastore is the pianist.

26. nepotism

(See No. 15.) By 1, neither Burr nor Fox holds the offices of president or treasurer and they are both office holders; therefore, X's are placed opposite each of the rows corresponding to Burr and Fox in the columns headed Pres., Tr., and N.O. (no officeholder). By 2, McCue is not president; therefore, X's are placed opposite McCue in the column headed Pres. and N.O. By 3, Burr is not the secretary. Therefore, an X is placed opposite Burr in the column headed Sec. It immediately follows from the table that Burr is the vice-president; therefore, an O is placed opposite Burr in the column headed V.P. and X's in the same column in all the other rows. Again, from the table Fox is the secretary; therefore, an O is placed opposite Fox in the column headed Sec. and X's in the same column in all the other rows. Then, from the table McCue is the treasurer; therefore, an O is placed opposite McCue in the column headed Tr. and X's in the same column in all the other rows. Now, either Marx or Palmer is the president. The first night, the treasurer, McCue; the vice-president, Burr; the secretary, Fox; and the president had dinner with the customer. But, by 5, Marx was not present at that dinner. Therefore, Marx is not the president. Thus, an X is placed opposite Marx in the column headed Pres. It then follows that Palmer is the president. Thus, an O is placed opposite Palmer in the column headed Pres. and an X in the same row in the column headed N.O. This leaves Marx as the member that holds no office. Hence, place an O opposite Marx in the column headed N.O. Therefore, Palmer is the president, Burr is the vice-president, Fox is the secretary, McCue is the treasurer, and Marx holds no office.

	Pres.	V.P.	Sec.	Tr.	N.O.
McCue	X	X	X	O	X
Fox	X	X	O	X	X
Burr	X	O	X	X	X
Palmer	O	X	X	X	X
Marx	X	X	X	X	O

27. sonny boy's letter

$$\begin{array}{r} S\ E\ N\ D \\ +\ M\ O\ R\ E \\ \hline M\ O\ N\ E\ Y \end{array}$$

Since this cryptogram represents an addition and each letter denotes a different digit, then the sum of the unit's column $D + E$ must be less than 20, so that we may carry at most 1. Similarly, the largest value N and R may have are 9 and 8 and, with the unit you may have carried, its sum should still be less than 20. This is true of $E + O$ and $S + M$; that is, $S + M$ must be less than 20. Hence, the M in the sum must represent the digit 1. Now $M = 1$ and $S + M = MO$; that is, $S + 1 = 10$. But, adding 1 to S, even if we have to carry 1 from $E + O$, gives at most 10 or 11. Hence, the letter O is either 1 or 0. Now, the letter O cannot be 1 for $M = 1$; hence, it must be zero. So far, we have result (1).

$$
(1) \qquad \begin{array}{r} S\ E\ N\ D \\ +\ 1\ 0\ R\ E \\ \hline 1\ 0\ N\ E\ Y \end{array}
$$

It follows that the letter S must be either 8 or 9. Assuming that $S = 8$, then $E + O$ must be greater than or equal to ten and this can only be, if $E = 9$ and $N + R$ is either equal to or greater than ten. But, if $E + O$ is either equal to or greater than ten, then $E + 0 + 1$ (carried) $= N$ and, since $E = 9$, then $E + 1 = 9 + 1 = 10$ and $N = 0$. But, N cannot equal 0, for the letter O equals zero. Hence, S cannot equal 8; therefore, $S = 9$. Since the letter O stands for the number 0, then in order to obtain $E + 0 = N$, we must have carried 1 from the sum $N + R$; that is, $E + 1 = N$. Now, adding the ten's column, we have $N + R = E$ or, since $N = E + 1$, $(E + 1) + R = E$, or $R + 1 = 0$. But, this is impossible, unless we had carried one unit of ten from the sum of the units $D + E$. Thus, $N + R + 1 = E + 10$; that is, $(E + 1) + R + 1 = E + 10$, $E + R + 2 = E + 10$, and $R = 8$. We can now write result (2).

$$\begin{array}{lr} & 9\ E\ N\ D \\ (2) & +\ 1\ 0\ 8\ E \\ \hline & 1\ 0\ N\ E\ Y \end{array}$$

Since in order to obtain the sum of the tens, $N + 8 = E$, we had to carry one unit of ten from the sum of the units $D + E$, then this sum is greater than or equal to 10. But, Y cannot be zero for the letter O is zero, hence, $D + E$ is greater than 10. We have already $S = 9$, $R = 8$, $M = 1$, and $O = 0$; hence, the only combinations whose sums are greater than 10 are $7 + 4$, $7 + 5$, $7 + 6$ and $6 + 5$. But, $7 + 4$ and $6 + 5$ would make $Y = 1$, but Y cannot be 1, for $M = 1$. Thus, the only possible sums are $7 + 5$ and $7 + 6$. Now E cannot be 7, for if it were, then $N = E + 1 = 7 + 1 = 8$, and N cannot be 8, for $R = 8$. Therefore, $D = 7$ and E must be either 5 or 6. But E cannot be 6, for if $E = 6$, then $N = E + 1 = 6 + 1 = 7$ and $D = 7$; hence, $E = 5$. It immediately follows that $Y = 2$ and $N = E + 1 = 5 + 1 = 6$. Thus, Sonny Boy asked his dad to send the amount of cents shown below in the solution (3), or $106.52.

$$\begin{array}{lr} & 9\ 5\ 6\ 7 \\ (3) & +\ 1\ 0\ 8\ 5 \\ \hline & 1\ 0\ 6\ 5\ 2 \end{array}$$

28. dad's answer to sonny boy's letter

Since the crytogram represents a subtraction, then in the hundred's column N which equals $E - E$ must

$$\begin{array}{r} S\ P\ E\ N\ D \\ -\ L\ E\ S\ S \\ \hline M\ O\ N\ E\ Y \end{array}$$

equal 0 or 9. That is, N = O if no borrowing is necessary, or N = 9 if we must borrow 1 unit of a thousand from P. But, if the remainder in the hundred's column is 9; that is, if N = 9, then the minuend N in the preceding column (ten's column) also equals 9. This minuend is then so large that no borrowing from P would be necessary. This gives us a contradiction. Hence, N cannot be 9; thus, N = 0. But, if N = 0, no borrowing from P is necessary. However, if the minuend N in the ten's column is 0, then we must borrow from the minuend E in the hundred's column no matter what S equals. But, if we borrow from the minuend E in the hundred's column, we must borrow from P to perform the subtraction E − E. This again leads to a contradiction. Thus, no matter which one of the possible values for N we choose, we get a contradiction and therefore the subtraction is impossible.

29. dad thinks it over and writes again

$$\begin{array}{r} S\ A\ V\ E \\ +\ M\ O\ R\ E \\ \hline M\ O\ N\ E\ Y \end{array}$$

Since the cryptogram represents an addition and each letter denotes a different digit, then the sum of the unit's column E + E must be at most 9 + 9 = 18, which is less than 20. Therefore, we may carry at most 1. Similarly, the largest value V and R could have are

9 and 8, and with the unit which may have been carried, the sum still would be less than 20. This is also true of A + O and S + M. That is, S + M must be less than 20; hence, the letter M in the sum must represent the digit 1. Now, M = 1 and S + M = MO; that is, S + 1 = 10. But, adding 1 to S, even if we have carried 1 from A + O, gives at most 10 or 11. Hence, the letter O is either 1 or 0. But the letter O cannot be 1, for M = 1; hence, it must be 0. Thus, so far we have result (1).

$$
(1) \qquad \begin{array}{r} S\ A\ V\ E \\ +\ 1\ 0\ R\ E \\ \hline 1\ 0\ N\ E\ Y \end{array}
$$

It follows that the letter S must be either 8 or 9. Assuming that S = 8, then A + 0 must be greater than or equal to 10. This can only be true, if A = 9 and V + R is either equal to or greater than ten. But, if A + 0 is greater than or equal to ten, then A + 0 + 1 (carried) = N and, since A = 9, then A + 0 + 1 = 9 + 1 = 10 and N = 0. But, N cannot equal 0, for O is equal to 0. Hence, S cannot equal 8; thus S = 9.

Since the letter O stands for 0, then in order to obtain A + 0 = N, we must have carried 1 from the sum V + R; that is, A + 1 = N. Now, since V + R is greater than 10, for E cannot equal zero, then E + E must be greater than 10 because Y cannot be zero, and we must carry one unit from the sum of the units E + E, so that E + E = 2E is greater than 10. But, if 2E is greater than 10, E must be greater than 5. Now we already have S = 9, M = 1, O = 0, and E must be greater than 5. Thus, in V + R = E, E must be greater than 5, so that the sum V + R must be greater than 15 and neither V nor R can equal 9 for S = 9. The only possible combination is 8 + 7. Assume that

$V = 7$, then $R = 8$, $E = 6$, and $Y = 2$; thus, so far, we have result (2).

$$\text{(2)} \qquad \begin{array}{r} 9\ A\ 7\ 6 \\ +\ 1\ 0\ 8\ 6 \\ \hline 1\ 0\ N\ 6\ 2 \end{array}$$

It follows that A may equal 5, 4, or 3. If $A = 5$, then, since $A + 1 = N$, $N = 6$. This is impossible for $E = 6$. If $A = 4$, then $N = A + 1 = 5$ and we obtain the sum 10562 shown in (3).

$$\text{(3)} \qquad \begin{array}{r} 9\ 4\ 7\ 6 \\ +\ 1\ 0\ 8\ 6 \\ \hline 1\ 0\ 5\ 6\ 2 \end{array}$$

If $A = 3$, then $N = A + 1 = 4$ and we obtain 10462 shown in (4).

$$\text{(4)} \qquad \begin{array}{r} 9\ 3\ 7\ 6 \\ +\ 1\ 0\ 8\ 6 \\ \hline 1\ 0\ 4\ 6\ 2 \end{array}$$

It is evident that if we assume $V = 8$ and $R = 7$, the sums will be the same as in the case where $V = 7$ and $R = 8$. Therefore, the amount sent by Dad to Sonny Boy is 10562 cents, or $105.62.

30. sonny boy goes on a business trip

Since this cryptogram represents an addition, each letter denotes a different digit; this problem is the same as No. 29. The reasoning necessary to solve this problem is exactly the same and may be duplicated by substituting T for S and K for V. The sums, as before, are 10562 and 10462.

31. a little business advice

$$\begin{array}{r} D\ E\ F\ E\ R \\ -\ D\ U\ T\ Y \\ \hline N\ O\ G\ O \end{array}$$

This cryptogram represents a subtraction. Since the answer does not contain a digit at the extreme left, then D must equal 1. Now, if $D = 1$, E must equal 1 or 0, for $10 + E - 1 = N$; that is, $9 + E = N$, and N cannot be greater than 9, for it is a digit. But E cannot equal 1, for $D = 1$; hence, $E = 0$. It immediately follows, since $9 + E = N$, that $N = 9$. Thus, so far, we have result (1).

$$(1) \qquad \begin{array}{r} 1\ 0\ F\ 0\ R \\ -\ 1\ U\ T\ Y \\ \hline 9\ O\ G\ O \end{array}$$

Now, to subtract T from 0, we must borrow one unit of a hundred from F; thus, $F - 1 = U + O$. Since we have already used 0, 1, and 9, and U and O are integers, then F may be 8, 7, or 6. If $F = 8$, then $U + O = 7$ and $U + O = 5 + 2$; if $F = 7$, then $U + O = 6$ and $U + O = 4 + 2$; if $F = 6$, then $U + O = 5$ and $U + O = 3 + 2$.

Assuming that $F = 8$, then let $U = 5$ and $O = 2$. We then have result (2).

$$\begin{array}{lr} & 1\ 0\ 8\ 0\ R \\ (2) & -\ 1\ 5\ T\ Y \\ \hline & 9\ 2\ G\ 2 \end{array}$$

From the ten's column, $10 - T = G$, or $10 = T + G$. Since we have already used 0, 1, 2, 5, 8, and 9, the possible combinations are $6 + 4$ and $7 + 3$. It may be easily shown that the combination $6 + 4$ is impossible; thus, let $T = 7$ and $G = 3$. We then have result (3).

$$\begin{array}{lr} & 1\ 0\ 8\ 0\ R \\ (3) & -\ 1\ 5\ 7\ Y \\ \hline & 9\ 2\ 3\ 2 \end{array}$$

Hence, $R - Y = 2$ and it immediately follows that $R = 6$ and $Y = 4$; thus, the operation is,

$$\begin{array}{r} 1\ 0\ 8\ 0\ 6 \\ -\ 1\ 5\ 7\ 4 \\ \hline 9\ 2\ 3\ 2 \end{array}$$

32. things are not what they seem

$$\begin{array}{r} F\ O\ U\ R \\ -\ T\ W\ O \\ \hline T\ E\ N \end{array}$$

This cryptogram represents a subtraction. Since the answer does not contain a digit at the extreme left, then F must equal 1. Now, if $F = 1$, from the hundred's column, we have $10 + O - T = T$, or $10 + O = 2T$;

it follows that O must represent an even digit; hence, O may assume the values, 2, 4, 6, or 8. Choosing the largest possible value for O, then O = 8; but 10 + O = 2T, therefore, 10 + 8 = 2T and T = 9. Thus, we have result (1).

$$\text{(1)} \qquad \begin{array}{r} 1\ 8\ U\ R \\ -\ 9\ W\ 8 \\ \hline 9\ E\ N \end{array}$$

So far, we have F = 1, O = 8, and T = 9. In order to carry out the subtraction in the unit's column; that is, R − 8 = N, we must borrow from U, for R cannot equal 9, since T = 9; thus, we have 10 + R − 8 = N, or R + 2 = N. Now, the greatest value N may have is 7. But, if N = 7, then R = 5 and we have result (2).

$$\text{(2)} \qquad \begin{array}{r} 1\ 8\ U\ 5 \\ -\ 9\ W\ 8 \\ \hline 9\ E\ 7 \end{array}$$

Since we have borrowed one unit from U and we cannot borrow from 8, we have U − 1 − W = E, or U − W − 1 = E. So far, we have T = 9, O = 8, N = 7, R = 5, and F = 1; hence, the greatest value U may have is 6. But, if U = 6, then W may equal 4, 3, or 2. Now, W cannot equal 4, for then E = U − W − 1 = 6 − 4 − 1 = 1 and F = 1; hence choose W = 3, then E = U − W − 1 = 6 − 3 − 1 = 2. Therefore, the operation is,

$$\begin{array}{r} 1\ 8\ 6\ 5 \\ -\ 9\ 3\ 8 \\ \hline 9\ 2\ 7 \end{array}$$

33. deceived by appearances

$$\begin{array}{r} T\ H\ R\ E\ E \\ -\ F\ O\ U\ R \\ \hline F\ I\ V\ E \end{array}$$

This cryptogram represents a subtraction. Since the answer does not contain a digit at the extreme left, then T must equal 1. Now, if $T = 1$, from the thousand's column, we have $10 + H - F = F$, or $10 + H = 2F$; hence, H may be any of the digits 2, 4, 6, or 8. However, before choosing a value for H, consider the other columns, in particular, the unit's column $E - R = E$. If we borrow a unit of 10 from E in the ten's column, then $10 + E - R = E$ and $10 = R$, which is impossible, because R is a digit. If we do not borrow, then $E - R = E$ and $R = 0$; therefore, $R = 0$. But if $R = 0$, then in the hundred's column we shall have $0 - O = 1$. To be able to perform this subtraction, we must borrow from H in the thousand's column. Hence, considering the thousand's column, we do not have $10 + H - F = F$, as stated below, but $10 + H - 1 - F = F$, or $9 + H = 2F$. Thus, the possible values of H are 1, 3, 5, 7, and 9. Now, H cannot be 9, for if $H = 9$, then $9 + H = 9 + 9 = 18 = 2F$ and $F = 9$ also. Therefore, the possible values for H are 1, 3, 5, or 7. If we let $H = 7$, so that $9 + H = 9 + 7 = 16 = 2F$, then $F = 8$ and, if we proceed to find a solution, we shall find that the operation is impossible; that is, H cannot equal 7. Now, let $H = 5$, then $9 + H = 9 + 5 = 14 = 2F$ and $F = 7$. Thus, so far we have result (1).

$$(1) \qquad \begin{array}{r} 1\ 5\ 0\ E\ E \\ -\ 7\ O\ U\ 0 \\ \hline 7\ I\ V\ E \end{array}$$

Consider now the unit's column E — 0 = E. Since E may have any value except 0, 1, 5, and 7, choose E = 9. We then have result (2).

$$(2)\qquad \begin{array}{r} 1\ 5\ 0\ 9\ 9 \\ -\ 7\ \mathrm{O}\ \mathrm{U}\ 0 \\ \hline 7\ \mathrm{I}\ \mathrm{V}\ 9 \end{array}$$

Thus, in the ten's column 9 — U = V. Since we have already used 0, 1, 5, 7, and 9, then U may have any of the values 2, 3, 4, 6, and 8. Now U cannot be 8 for 9 — U = 9 — 8 = 1 = V and T = 1. Nor can U equal 4, because 9 — U = 9 — 4 = 5 = V and H = 5; thus, the only possible combination is 6 and 3. Let U = 6, then 9 — U = 9 — 6 = 3 = V to obtain result (3).

$$(3)\qquad \begin{array}{r} 1\ 5\ 0\ 9\ 9 \\ -\ 7\ \mathrm{O}\ 6\ 0 \\ \hline 7\ \mathrm{I}\ 3\ 9 \end{array}$$

Now consider the hundred's column 0 — O = I. To perform this operation, we must borrow from 5 in the thousand's column, so that the operation become 10 + 0 — O = I, or simply 10 — O = I. We have already used 0, 1, 3, 5, 6, 7, and 9; thus, the only possible combination is 8 and 2. Hence, choose O = 8 and I = 2 to obtain the correct solution,

$$\begin{array}{r} 1\ 5\ 0\ 9\ 9 \\ -\ 7\ 8\ 6\ 0 \\ \hline 7\ 2\ 3\ 9 \end{array}$$

34. the cat

```
lines 1.              C A T
      2.    R C ) A P D T M
      3.        A D C
      4.          N N T
      5.            B A
      6.            A E M
      7.            A E M
```

From lines 1, 2, and 3, the product of RC by C is a number which ends in C; thus, C must be 1, 5, or 6. Since the product of RC by C is a number of three digits, then C cannot be 1; hence, C is either 5 or 6. Suppose C = 5, then from lines 1, 2, and 5, RC × A = BA; that is, A × C, or A × 5 equals a number ending in A. But any number times 5 ends in 0 or 5, so that A would have to be 0 or 5; but, A cannot be 5 for C = 5; hence, A would have to be 0. From lines 4, 5, and 6, we have T − A = E and, if A = 0, then T = E, which is impossible. Hence, A cannot be 0; therefore, C cannot be 5; hence, C = 6.

From lines 2, 3, and 4, if C = 6, we have,

```
A P D
A D 6
-----
  N N
```

Now, D cannot equal 6, for C = 6; thus, D is either less than or greater than 6. If D is less than 6, we must borrow from P, so that,

$$\begin{array}{ccc} A & P-1 & D+10 \\ A & D & 6 \\ \hline & N & N \end{array}$$

It follows that $D + 10 - 6 = N$, or $D + 4 = N$, and that $P - 1 - D = N$; hence, $D + 4 = P - 1 - D$, or $2D + 5 = P$. Since P must be a digit, then D must be less than 3; that is, D may be equal to 0, 1, or 2.

From lines 1, 2, and 3, $RC \times C = ADC$; that is $R6 \times 6 = AD6$, or R times $6 + 3 = AD$, a number ending in D. It immediately follows that no matter what digit R represents, D cannot be 0. Moreover, D cannot equal 2, for no matter what digit R represents, $R \times 6 + 3$ is an odd number; thus, $D = 1$. If $D = 1$, R must be 3 or 8, for $R \times 6 + 3$ must be a number ending in D; that is, 1. If $D = 1$ and $R = 3$, then $RC \times C = 36 \times 6 = 216 = ADC$; thus, $A = 2$. Moreover, since $D + 4 = N$, then $N = 1 + 4 = 5$. (See above.) Then from lines 1, 2, and 5, $RC \times A = 36 \times 2 = 72 = BA$. But from lines 4, 5, and 6,

$$\begin{array}{ccc} N\ N\ T & & 5\ 5\ T \\ B\ A & \text{or} & 7\ 2 \\ \hline A\ E & & 2\ E \end{array}$$

Now, no matter what digit T represents, the difference $55T - 72$ equals a three-digit number and 2E represents a two-digit number. Hence, D cannot equal 1; therefore, D cannot be less than 6. If D is greater than 6, then D may be 7, 8, or 9. Again, from lines 1, 2, and 3, $RC \times C = ADC$; that is $R6 \times 6 = AD6$. Hence, $R \times 6 + 3 = AD$. This number ends in D; hence, it must end in 7, 8, or 9. No matter what digit R represents, $R \times 6 + 3$ is an odd number; thus, D cannot be 8, so D must be 7 or 9. The only possible values for R are then 4 or 9. But, if $R = 9$, then $RC \times C = 96 \times 6 = 576 = ADC$, $A = 5$, and $D = 7$. Now, from lines 1, 2, and 5, $RC \times A = BA$ is a two-digit number, but $RC \times A = 96 \times 5 = 480$ is a three-digit number. Hence, R cannot be 9; therefore, $R = 4$.

From lines 1, 2, and 3, $RC \times C = 46 \times 6 = 276 = ADC$; therefore, $A = 2$ and $D = 7$. Moreover, from lines 2, 3, and 4,

$$\begin{array}{ccc} A & P & D \\ A & D & C \\ \hline & N & N \end{array} \quad \text{or} \quad \begin{array}{ccc} 2 & P & 7 \\ 2 & 7 & 6 \\ \hline & N & N \end{array}$$

Hence, $N = 7 - 6 = 1$ and $P = N + 7 = 1 + 7 = 8$. So far, we have the result shown below.

$$\begin{array}{lrrrrrrr}
\text{lines } 1. & & & & & 6 & 2 & T \\
2. & 4 & 6 &)2 & 8 & 7 & T & M \\
3. & & & 2 & 7 & 6 & & \\
4. & & & & 1 & 1 & T & \\
5. & & & & & B & 2 & \\
6. & & & & & 2 & E & M \\
7. & & & & & 2 & E & M
\end{array}$$

From lines 1, 2, and 5, $46 \times 2 = 92 = B2$; hence, $B = 9$. Then, from lines 4, 5, and 6, T must be greater than 2, for $N = 1$ and $A = 2$ and there is no carry since $B = 9$. Therefore, T may equal 3 or 5, for we have already used 4, 6, 7, 8, and 9. If $T = 3$, then $T - 2 = 3 - 2 = 1 = E$, which is impossible, for $N = 1$; hence, $T = 5$ and $E = T - 2 = 5 - 2 = 3$. Finally, from lines

1, 2, and 6, $46 \times T = 46 \times 5 = 230 = 2EM$; therefore, $M = 0$. Thus, the complete solution is,

```
        6 2 5
     ---------
  4 6)2 8 7 5 0
      2 7 6
      -----
        1 1 5
          9 2
        -----
          2 3 0
          2 3 0
          -----
```

35. the arab

```
lines 1.                  A R A B
                  -----------------
      2. P R E)A N M M R D R
      3.         P R E
               -------
      4.         C S R R
      5.         C C P R
                 -------
      6.           A E M D
      7.             P R E
                   -------
      8.             D E A R
      9.             D E A R
                     -------
```

From lines 4, 5, and 6, we have $R - R = M$; hence, $M = 0$. From lines 1, 2, and 3, $A \times PRE = PRE$; thus, $A = 1$. From lines 2, 3, and 4,

```
A N M M
  P R E
-------
  C S R
```

Substituting $A = 1$ and $M = 0$,

$$\begin{array}{rrrr} 1 & N & 0 & 0 \\ & P & R & E \\ \hline & C & S & R \end{array}$$

To subtract the unit's digits $0 - E$, we must borrow from 0 in the ten's column, but we cannot do this unless we borrow from N in the hundred's column; thus,

$$\begin{array}{rrrr} 1 & N-1 & 9 & 10 \\ & P & R & E \\ \hline & C & S & R \end{array}$$

It follows that $10 - E = R$ and $9 - R = S$, or $9 - S = R$, so that $10 - E = 9 - S$, or $S = E - 1$. Substituting $M = 0$ and $A = 1$ in lines 6, 7, and 8 gives,

$$\begin{array}{rrrr} 1 & E & 0 & D \\ & P & R & E \\ \hline & D & E & 1 \end{array}$$

Thus, $D - E = 1$, or $D = E + 1$. To subtract $0 - R$ in the ten's column, we must borrow from E in the hundred's column, so that $10 + (E - 1) - P = D$, or $9 + E - P = D$, but $D = E + 1$; hence, $9 + E - P = E + 1$ and $P = 8$.

Now consider lines 4, 5, and 6,

$$\begin{array}{rrrr} C & S & R & R \\ C & C & P & R \\ \hline & A & E & M \end{array} \quad \text{or} \quad \begin{array}{rrrr} C & S & R & R \\ C & C & 8 & R \\ \hline & 1 & E & 0 \end{array}$$

Then, $R - 8 = E$, but R cannot equal 9, for from lines 6, 7, and 8; $10 - R = E$. If $R = 9$, then $10 - R = 10 - 9 = 1 = E$, which is impossible, because $A = 1$. Hence, R must be less than 8. To perform the operation $R - 8$, we must borrow from S in the hundred's column; thus, $10 + R - 8 = 2 + R = E$. But, from lines 6, 7, and 8, we have $10 - R = E$; hence, $10 - R = 2 + R$, $8 = 2R$, and $R = 4$. It follows that $E = 10 - R = 10 - 4 = 6$, $D = E + 1 = 6 + 1 = 7$, and $S = E - 1 = 6 - 1 = 5$. The result is shown below.

```
lines 1.                 1 4 1 B
                    _____________
      2. 8 4 6 )1 N 0 0 4 7 4
      3.          8 4 6
                 -------
      4.          C 5 4 4
      5.          C C 8 4
                  -------
      6.            1 6 0 7
      7.              8 4 6
                    -------
      8.              7 6 1 4
      9.              7 6 1 4
                      -------
```

Now, from lines 2, 3, and 4,

```
1 N 0 0
  8 4 6
-------
  C 5 4
```

So that $10 + N - 1 - 8 = C$, or $1 + N = C$. Then, from lines 4, 5, and 6,

```
  C 5 4 4
  C C 8 4
---------
    1 6 0
```

Thus, from the hundred's column, $4 - C = 1$; hence, $C = 3$. But, $1 + N = C$, therefore, $1 + N = 3$; thus, $N = 2$. Finally, from lines 1, 2, and 8, $846 \times B = 7614$; hence, $B = 7614 \div 846 = 9$. Thus, the complete solution is,

```
              1 4 1 9
      ______________
8 4 6)1 2 0 0 4 7 4
        8 4 6
      -------
        3 5 4 4
        3 3 8 4
        -------
          1 6 0 7
            8 4 6
          -------
            7 6 1 4
            7 6 1 4
            -------
```

36. a good one

```
line 1.                   G O O D O N E
                       ________________
     2.   P R R)D R E O O O D M N
     3.         D R M
                -----
     4.             P O O O
     5.               R R D
                    -------
     6.                   D D M
     7.                   G R E
                          -----
     8.                   P D R N
     9.                   P D R N
                          -------
```

From lines 2, 3, and 4, we have,

$$\begin{array}{ccc} D & R & E \\ D & R & M \\ \hline & & P \end{array}$$

Bringing down the next digit O, we obtain PO, a two-digit number. Now, the divisor PRR is a three-digit number, hence it does not go into the two-digit number PO. Therefore, the second digit in the quotient; that is, O, must equal 0. From lines 4, 5, and 6,

$$\begin{array}{cccc} P & O & O & O \\ & R & R & D \\ \hline & & & D \end{array} \quad \text{or} \quad \begin{array}{cccc} P & 0 & 0 & 0 \\ & R & R & D \\ \hline & & & D \end{array}$$

To subtract the unit's digit D from 0, we must borrow from the 0 in the ten's place; to subtract the ten's digit R from 0, we must borrow from the 0 in the hundred's place; to subtract the hundred's digit R from 0, we must borrow from P in the thousand's place. Thus, we have,

$$\begin{array}{cccc} P-1 & 9 & 9 & 10 \\ & R & R & D \\ \hline & & & D \end{array}$$

It follows that $10 - D = D$, or $10 = 2D$, and $D = 5$. Moreover, $9 - R = 0$, so $R = 9$ and, since $P - 1 = 0$, then $P = 1$. Thus, so far, we have the result shown below.

lines										
1.				G	0	0	5	0	N	E
2.	1 9 9)	5	9	E	0	0	0	5	M	N
3.		5	9	M						
4.				1	0	0	0			
5.					9	9	5			
6.							5	5	M	
7.							G	9	E	
8.							1	5	9	N
9.							1	5	9	N

From lines 6, 7, and 8,

$$\begin{array}{ccc} 5 & 5 & M \\ G & 9 & E \\ \hline 1 & 5 & 9 \end{array}$$

Neither M nor E can be equal to 0 for O = 0; thus, M must be less than E for M = E + 9, which is impossible if M is a digit. To subtract E from M, we must borrow from the 5 in the ten's place and, in order to perform the subtraction in the ten's column, we must borrow from the 5 in the hundred's column; hence, we have,

$$\begin{array}{ccc} 4 & 4 & M+10 \\ G & 9 & E \\ \hline 1 & 5 & 9 \end{array}$$

It follows that $4 - G = 1$ and $G = 3$; thus, $M + 10 - E = 9$ and $M + 1 = E$. Now, from lines 1, 2, and 3, $199 \times G = 59M$. Since $G = 3$, then $199 \times 3 = 597 = 59M$ and $M = 7$, but $M + 1 = E$; hence, $7 + 1 = E = 8$. From lines 1, 2, and 7, $199 \times N = G9E$. Since $G = 3$ and $E = 8$, then $199 \times N = 398$ or $N = 398 \div 199 = 2$.

Thus, the complete solution is,

```
           3 0 0 5 0 2 8
      ___________________
1 9 9)5 9 8 0 0 0 5 7 2
      5 9 7
      -----
          1 0 0 0
            9 9 5
          -------
              5 5 7
              3 9 8
              -----
              1 5 9 2
              1 5 9 2
              -------
```

37. the three fives

Rewrite the blurred example as follows,

```
          a b c d e f g h i     j k l     m n o p q r s
lines A.  5 * * * * * * * *  ÷  * * *  =  * * * * * * *
      B.  * * *
          -----
      C.      * * * *
      D.        * * *
              -------
      E.          5 * *
      F.          * * *
                  -----
      G.          * 5 * *
      H.          * * * *
                  -------
```

From line C, we observe that three digits must be brought down from the dividend before the three-digit divisor jkl will be contained in this partial remainder. Hence, the digits An and Ao, or simply n and o, in line A in the quotient must equal 0.

From lines A, B, and C, we obtain,

$$\begin{array}{rr} \text{lines A.} & 5\ *\ * \\ \text{B.} & *\ *\ * \\ \hline \text{C.} & * \end{array}$$

That is, the difference between these two three-digit numbers is a number of one digit; therefore, the digit Ba = 5. Line E shows that we must bring down two digits from the dividend before the divisor jkl is contained in this partial remainder. But, line D is jkl $\times$ p, for the digits n and o in the quotient equal 0. Hence, Aq or simply q = 0.

From lines C, D, and E,

$$\begin{array}{rr} \text{lines C.} & *\ *\ *\ * \\ \text{D.} & *\ *\ * \\ \hline \text{E.} & 5 \end{array}$$

This 5 is the difference between a four-digit number and a three-digit number, or the sum of a three-digit number, and 5 equals a four-digit number. But, adding 5 to a three-digit number will not result in a four-digit number unless the ten's and hundred's digits in the three-digit number equal 9; thus, digits Dd and De both equal 9. Moreover, if both the ten's and hundred's digits in line D equal 9, then the thousand's digit in line C must equal 1, and the hundred's and ten's digits are both equal to 0; that is, Cc = 1 and Cd = Ad = Ce = Ae = 0.

From line D, jkl $\times$ p = 99* or $p = \frac{99*}{jkl}$. Now, from line B, jkl $\times$ m = 5** or $m = \frac{5**}{jkl}$, so that jkl lies between 100 and 599; hence $p = \frac{99*}{jkl}$ must lie between 1 and 9.

If $p = 1$, then $jkl = 99*$, but jkl must be less than 599; hence, p cannot equal 1.

If $p = 2$, then $jkl = \frac{99*}{2} = 49*$; but, from line B, $jkl \times m = 5**$, or $m = \frac{5**}{jkl} = \frac{5**}{49*}$, and there is no digit $m = \frac{5**}{49*}$. Hence, p cannot equal 2. Similarly, p cannot equal 3 or 4, for if $p = 3$, then $m = \frac{5**}{33*}$, and, if $p = 4$, then $m = \frac{5**}{24*}$; but, this is impossible.

If $p = 6$, then $jkl = \frac{99*}{6} = 16*$ and $m = \frac{5**}{16*}$, so that m could equal 3 and jkl would then have to be a number between 167 and 169. But, $jkl \times p = 99*$ and the numbers 167, 168, or 169 multiplied by 6 yields a product greater than 1000; hence, p does not equal 6.

If $p = 7$, then $jkl = \frac{99*}{7} = 14*$ and $m = \frac{5**}{14*}$; thus, m could be 4 and jkl would then have to be a number between 140 and 149. But, $143 \times 7 = 1001$, so jkl must lie between 140 and 142. Again line $D = jkl \times 7 = 99*$ and 140 and 141×7 yields a product less than $141 \times 7 = 987$, so that jkl would have to be 142. The line $D = 142 \times 7 = 994$. But, line C = line D + 5 and $994 + 5 = 999$, which is less than 1000; hence, p cannot equal 7.

If $p = 8$, then $jkl = \frac{99*}{8} = 12*$ and $m = \frac{5**}{12*}$; thus, m could be 4 and jkl would have to be a number between 127 and 129. But, any number between 127 and 129 times 8 would yield a number greater than or equal to $127 \times 8 = 1016$ and $jkl \times 8 = 99*$; therefore, p cannot be 8.

If $p = 9$, then $jkl = \frac{99*}{9} = 11*$ and $m = \frac{5**}{11*}$, so that m could be 5. Then jkl would have to be a number between 111 and 119. Now, any number greater than or equal to 112 would yield a product greater than or equal to $112 \times 9 = 1008$, so jkl would have to be 111.

Line H = jkl × s and, if jkl = 111, then jkl × s = 111 × s. No matter what value the digit s may take it cannot be greater than 111 × 9 = 999, and line H is a four-digit number. Therefore, p cannot equal 9.

Finally, if p = 5, then jkl = $\frac{99*}{5}$ = 19* and m = $\frac{5**}{19*}$, so that m could be 3; then jkl would have to be a number between 190 and 199, so that jkl would have to be 198 or 199 for jkl × 5 = 99*. Now, if jkl = 198, then line D = jkl × p = 198 × 5 = 990. But, line C = line D + line E, or 100* = 990 + 5, which is impossible. Therefore, jkl = 199, for 199 × 5 = 995 and 995 + 5 = 1000 = line C. That is, Cf = Af = 0, Df = 5, and p = 5; hence, jkl = 199, Aj = 1, Ak = 9, and Ai = 9. Thus, so far we have the following results,

	a	b	c	d	e	f	g	h	i		j	k	l		m	n	o	p	q	r	s
lines A.	5	*	*	0	0	0	*	*	*	÷	1	9	9	=	*	0	0	5	0	*	*
B.	5	*	*																		
C.			1	0	0	0															
D.				9	9	5															
E.						5	*	*													
F.						*	*	*													
G.						*	5	*	*												
H.						*	*	*	*												

Line B = jkl × m = 199 × m = 5***; thus, m must be 3 and line B = 199 × 3 = 597; that is, Bb = 9 and Bc = 7. Since line A − line B = 1, then line A = line B + 1 = 597 + 1 = 598 and Ab = Bb = 9, Ac = 8, Bc = 7, and m = 3. Line F = jkl × r = 199 × r = ***, which is a three-digit number less than line E = 5**; that is, less than 600. Now, 199 × 4 = 796 and 199 × 3 = 597, so that r may equal 1, 2, or 3. If r = 3, than jkl $\times$ r = 199 $\times$ 3 = 957; but, from lines E, F, and G, we have,

$$\begin{array}{ll} \text{E.} & 5 * * \\ \text{F.} & * * * \\ \hline \text{G.} & * 5 * \end{array}$$

Then the difference, line G, has the extreme left-hand digit different from 0; that is, the difference is a three-digit number, while the difference 5** $-$ 597 must be a number of one digit. Hence, r cannot equal 3.

Assume that r = 1, then jkl $\times$ r = 199 $\times$ 1 = 199 and the difference 5** $-$ 199 is evidently greater than 199. But, the partial remainder, line G, must be less than the divisor 199; hence, r cannot equal 1. Therefore, r = 2 and it follows that jkl $\times$ r = 199 $\times$ 2 = 398; that is, Ff = 3, Fg = 9, and Fh = 8. Thus, the difference line G = line E $-$ line F which becomes,

$$\begin{array}{ll} \text{E.} & 5 * * \\ \text{F.} & 3\ 9\ 8 \\ \hline \text{G.} & * 5 * \end{array}$$

Line G must be less than 199, so the hundred's digit in line G must be 1; that is, Gf = 1. It immediately follows that the ten's digit in line E must be 4, if there is no borrowing, and 5, if we must borrow to subtract the unit's digit. Since the final remainder is zero, then line H = line G, so that Hf = 1 and Hg = 5. But, line H = jkl $\times$ s; that is, 199 $\times$ s = 15**, a four-digit number, equals 1500 plus **. Hence, s must equal 8, and jkl $\times$ s = 199 $\times$ 8 = 1592 = line H = line G. Therefore, Gh = Hh = 9 and Gi = Hi = Ai = 2. But, line E $-$ line F = line G, or line E = line F + line G = 398 + 159 = 557; hence, Eg = Ag = 5 and Eh = Ah = 7. Therefore, the complete solution is,

```
           3 0 0 5 0 2 8
     ___________________
1 9 9)5 9 8 0 0 0 5 7 2
      5 9 7
      -----
          1 0 0 0
            9 9 5
          -------
              5 5 7
              3 9 8
              -----
              1 5 9 2
              1 5 9 2
              -------
```

38. the five fours

Rewrite the blurred example as follows:

```
          a b c d e f g   h i j   k l m n
lines A.  * * * * * * 4 ÷ * * * = * 4 * *
      B.    * * *
          ---------
      C.    * * 4 *
      D.    * * * 4
            -------
      E.      * * * *
      F.        * 4 *
              -------
      G.        * * * *
      H.        * * * *
                -------
```

It is easily seen that Gg = Hg = 4. From lines A, B, and C,

```
A.  * * * *
B.    * * *
    -------
C.    * * 4
```

That is, line B + line C = line A. Since lines B and C are each a three-digit number and line A is a four-digit number, then the thousand's digit in line A must be 1, so that Aa = 1. Similarly, line G + line F = line E and lines G and F are both three-digit numbers (before bringing down Gg) while line E consists of four digits; hence, Ec = 1. Now line D = hij × 4 = ***4, a four-digit number, while line B = hij × k = ***, a three-digit number, and line F = hij × m = ***, a three-digit number; therefore, both k and m are less than 4. Moreover, hij × 4 = ***4, a number ending in 4; that is, j × 4 must equal a number ending in 4. Hence, j equals 1 or 6.

Assume that j = 1, then line H = ***4 = hij × n, so that if j = 1, then n = 4, since j × n = 4; thus, line F = hij × m = *4*. Now, m is less than 4, so that m equals 1, 2, or 3. Hence, if J = 1, n = 4, m = 1, line F = hij × m = hij × 1 = *4*, i = 4, and Ff = 1, so that hij = h41. Consequently, line H = line D = h41 × 4 = (4h + 1) 64†, so that Hf = Gf = Dd = 6. Then, from lines C, D, and E,

```
C.  * * 4 *
D.  * * 6 4
    -------
E.    1 * *
```

No matter what the unit's digit in line C may be, the ten's digit in line E will have to be either 8 or 7. This means that the hundred's digit in line F will have to be 9 or 8, but line F = hij × m = h41 × 1 = 94*, or 84*, so that h must be 9 or 8. But, if h = 9, line D = 941 × 4 = 3764 = line H = line G and if h = 8, line D = 841 × 4 = 3364 = line H = line G. From lines E, F, and G, we then have,

```
1 8 * *        1 7 * *
  9 4 1   or     8 4 1
-------        -------
  3 7 6          3 7 6
```

† (4h + 1) is the hundred's digit—the parentheses do not indicate multiplication.

But, this is impossible; therefore, m cannot equal 1.

If $j = 1$, $n = 4$, and $m = 2$, then line F $=$ hij $\times$ m $=$ hi1 $\times$ 2 $=$ *4*, so that $2i = 4$, $i = 2$, and Ff $= 2$; hence, hij $=$ h21. Now, line H $=$ line D $=$ h21 $\times$ 4 $=$ $(\underline{4h})\underline{84}$, so that Hf $=$ Gf $=$ Dd $= 8$. Then, from lines C, D, and E,

$$\begin{array}{lrrrr} \text{C.} & * & * & 4 & * \\ \text{D.} & * & * & 8 & 4 \\ \hline \text{E.} & & 1 & * & * \end{array}$$

No matter what the unit's digit in line C may be, the ten's digit in line E will have to be 7 or 6. This means that the hundred's digit in line F will have to be 8 or 7, but line F $=$ hij $\times$ m $=$ h21 $\times$ 2 $=$ 842 or 742. Line F cannot be equal to 742, so that h must be 8. But, if $h = 8$, line D $= 842 \times 4 = 3368 =$ line H $=$ line G, then from lines E, F, and G,

$$\begin{array}{lrrrr} \text{E.} & 1 & 7 & * & * \\ \text{F.} & & 8 & 4 & 2 \\ \hline \text{G.} & & 3 & 3 & 6 \end{array}$$

But, this is impossible; therefore, m cannot equal 2. If $j = 1$, $n = 4$, and $m = 3$, then line F $=$ hij $\times$ m $=$ hi1 $\times$ 3 $=$ *4*, so that $3i = 4$ and i cannot be a digit. Hence, m cannot equal 3, and, since m cannot equal 1 or 3, then j cannot equal 1 and n cannot equal 4.

If $j = 6$, line H $=$ ***4 $=$ hij $\times$ n $=$ hi6 $\times$ n, so that n may equal 4 or 9, for $6 \times n$ must be a number ending in 4. Assume that $j = 6$ and $n = 4$, then, again, since m must be less than 4, m may equal 1, 2, or 3. If $j = 6$, $n = 4$, and $m = 1$, then line F $=$ hij $\times$ m $=$ hi6 $\times$ 1 $=$ *4*, $i = 4$, and Ff $= 6$, so that hij $=$ h46. Then, line H $=$

line D = h46 × 4 = $(\underline{4h + 1})\underline{84}$, so that Hf = Gf = Dd = 8. Thus, from lines C, D, and E,

```
C.  * * 4 *
D.  * * 8 4
   ---------
E.    1 * *
```

No matter what the unit's digit in line C may be, the ten's digit in line E will have to be 6 or 5. This means that the hundred's digit in line F will have to be 7 or 6. But, line F = hij × m = h46 × 1 = 746, or 646. So that h must be 7 or 6. If h = 7 or 6, line D = 746 × 4 = 2984, or line D = 646 × 4 = 2584; then, lines E, F, and G would be,

```
1 7 * *          1 6 * *
  7 4 6    or      6 4 6
-------          -------
  2 9 8            2 5 8
```

This is impossible; therefore, m cannot equal 1.

Let j = 6, n = 4, and m = 2, then line F = hij × m = hi6 × 2 = *4* and 2i + 1 = 4, or 2i = 3, but there is no digit i such that 2i = 3; hence, m cannot equal 2. Now, assume that j = 6, n = 4, and m = 3, then line F = hij × m = hi6 × 3 = *4* and 3i + 1 = 4, or 3i = 3, and i = 1, Ff = 8 and hij = h16. Then, line H = line D = hi6 × 4 = $(\underline{4h})\underline{64}$, so that Hf = Gf = Dd = 6. Then, from lines C, D, and E,

```
C.  * * 4 *
D.  * * 6 4
   ---------
E.    1 * *
```

No matter what the unit's digit in line C may be, the ten's digit in line E will have to be 8 or 7. This means that the hundred's digit in line F will have to be 9 or 8. But, line F $= hij \times m = h16 \times 3 = 948$ or 848, so that $3h = 9$ and $h = 3$, or $3h = 8$, which is impossible. Thus, $hij = 316$ and line D $= 316 \times 4 = 1264$. Then, from lines E, F, and G,

$$\begin{array}{r} 1\ 8\ *\ * \\ 9\ 4\ 8 \\ \hline 1\ 2\ 6 \end{array}$$

However, this is impossible; therefore, m cannot equal 3 and since m cannot equal 1, 2, or 3, then $j = 6$ and $n = 4$ is impossible.

Let $j = 6$ and $n = 9$ and since m is equal to 1, 2, or 3, consider first $j = 6$, $n = 9$, and $m = 1$. Line F $= hij \times m = hi6 \times 1 = *4*$, so that $i = 4$ and $hij = h46$. Line H $= hij \times n = h46 \times 9 = (\underline{9h + 4})\underline{14} = **14$, so that $Hf = Gf = 1$. Line B $= hij \times k$; now, k must be less than 4; that is, k may equal 1, 2, or 3.

Assume that $j = 6$, $n = 9$, $m = 1$, and $k = 3$, then line B $= hij \times k = h46 \times 3 = (\underline{3h + 1})\underline{38} = ***$, a three-digit number; hence, h must equal 1 or 2. If $h = 1$, then line D $= hij \times 4 = 146 \times 4 = 584$, but line D is a four-digit number; hence, h cannot equal 1. If $h = 2$, line D $= 246 \times 4 = 984$, but line D is a four-digit number; hence, h cannot equal 2 and, therefore k cannot equal 3.

Let $j = 6$, $n = 9$, $m = 1$, $k = 2$, and line B $= hij \times k = h46 \times 2 = (\underline{2h})\underline{92}$; then, h may equal any number from 1 to 4. Now, line D $= hij \times 4 = h46 \times 4 = ***4$, a four-digit number and $h46 \times 4 = (\underline{4h + 1})\underline{84}$. It follows h cannot be 1 or 2, for then line D would be a three-digit number; thus, h is either 3 or 4. If $k = 2$ and $h = 3$, line D $= 346 \times 4 = 1384$; then, from lines C, D, and E,

$$\begin{array}{lr} \text{C.} & * * 4 * \\ \text{D.} & 1\,3\,8\,4 \\ \hline \text{E.} & 1 * * \end{array}$$

Thus, the thousand's digit in line C must be 1; that is, Cb = 1 and the hundred's digit in line C must be 5. It follows from lines A, B, and C that,

$$\begin{array}{lrl} \text{A.} & 1 * * * & \\ \text{B.} & 6\,9\,2 & = 346 \times 2 = \text{hij} \times \text{k} \\ \cline{2-2} \text{C.} & 1\,5\,4 & \end{array}$$

This is impossible; hence, h cannot equal 3, so it must equal 4.

Let k = 2 and h = 4, then hij = 446 and line D = hij × 4 = 446 × 4 = 1784. From lines C, D, and E, we then have,

$$\begin{array}{lr} \text{C.} & * * 4 * \\ \text{D.} & 1\,7\,8\,4 \\ \hline \text{E.} & 1 * * \end{array}$$

Thus, Cb = 1, Cc = 9, and Ed = 6. Line F = hij × m = 446 × m = *4*, so that m = 1, Fd = 4, and Ff = 6. Then from lines E, F, and G,

$$\begin{array}{lrl} \text{E.} & 1\,6 * * & \\ \text{F.} & 4\,4\,6 & \\ \cline{2-2} \text{G.} & 1\,2 * * & \text{or} \quad 1\,1 * * \quad \text{(a four-digit number)} \end{array}$$

The difference between lines E and F is a three-digit number; hence, h cannot equal 3 and, since h cannot be 1, 2, 3, or 4, then k cannot equal 2; therefore, k must equal 1. Now, assume that j = 6, n = 9, and k = 1; then, line D = hij × 4 = h46 × 4 = $\underline{(4h + 1)}\,\underline{84}$ = ***4. Thus, h must be greater than 2.

If h = 3, line F = hij × m = 346 × m = *4*, so that m = 1, Ff = 6 and Fd = 3. Then line F = 346. But line G = line H = hij × n = 346 × 9 = 3114.

From lines E, F, and G, we obtain,

$$\begin{array}{lr} \text{E.} & 1 * * * \\ \text{F.} & 3\ 4\ 6 \\ \hline \text{G.} & 3\ 1\ 1 \end{array}$$

But, 346 + 311 = 657, which is less than 1000; hence, h cannot equal 3.

If h = 4, line F = hij × m = 446 × m = *4*, so that m = 1, Ff = 6, and Fd = 4; thus, line F = 446. But line G = line H = hij × n = 446 × 9 = 4014, so that from lines E, F, and G,

$$\begin{array}{lr} \text{E.} & 1 * * * \\ \text{F.} & 4\ 4\ 6 \\ \hline \text{G.} & 4\ 0\ 1 \end{array}$$

Since 446 + 401 = 847, which is less than 100, h cannot equal 4.

If h = 5, line F = hij × m = 546 × m = *4*, so that m = 1, Ff = 6, and Fd = 5; hence, line F = 546. But line G = line H = hij × n = 546 × 9 = 4914, so that from lines E, F, and G,

$$\begin{array}{lr} \text{E.} & 1 * * * \\ \text{F.} & 5\ 4\ 6 \\ \hline \text{G.} & 4\ 9\ 1 \end{array}$$

Hence, line E = 546 + 491 = 1047; that is, Ef = 7, Ee = 3, Ed = 0, and Ec = 1. Then, from lines C, D, and E,

```
C.  * * 4 *
D.  2 1 8 4 = 546 × 4 = hij × 4
    -------
E.    1 0 3
```

The ten's column is 14 − 8; or 13 − 8; that is, 6 or 5, and does not equal 0. Hence, h cannot equal 5.

If h = 6, line F = hij × m = 646 × m = *4*, so that m = 1 and line F = 646; but, line G = line H = hij × n = 646 × 9 = 5814. Thus, from lines E, F, and G, line E = line F + line G = 646 + 581 = 1227; that is, Ef = 7, Ee = 2, Ed = 2, and Ec = 1. Then, from lines C, D, and E,

```
C.  * * 4 *
D.  2 5 8 4 = 646 × 4 = hij × 4
    -------
E.    1 2 2
```

The ten's column is 14 − 8 or 13 − 8; that is, 6 or 5, which does not equal 2. Hence, h cannot equal 6.

If h = 7, line F = hij × m = 746 × m = *4*, so that m = 1 and line F = 746. Now, line G = line H = hij × n = 746 × 9 = 6714, so that from lines E, F, and G, line E = line F + line G = 746 + 671 = 1417; that is, Ef = 7, Ee = 1, Ed = 4, and Ec = 1. Then, from lines C, D, and E,

```
C.  * * 4 *
D.  2 9 8 4 = 746 × 4 = hij × 4
    -------
E.    1 4 1
```

The ten's column is 6 or 5 and does not equal 4; hence, h cannot be 7.

If h = 9, line F = hij × m = 946 × m = *4*, so that m = 1 and line F = 946. But then, line G = line H =

hij $\times$ n = 946 $\times$ 9 = 8514, so that from lines E, F, and G, line E = line F + line G = 946 + 851 = 1797; that is, Ef = 7, Ee = 9, Ed = 7, and Ec = 1. Then from lines C, D, and E,

$$\begin{array}{ll} \text{C.} & * \ * \ 4 \ * \\ \text{D.} & \underline{3\ 7\ 8\ 4} = 946 \times 4 = \text{hij} \times 4 \\ \text{E.} & \quad 1\ 7\ 9 \end{array}$$

The ten's column is 6 or 5 and does not equal 7; hence, h cannot equal 9.

If h = 8, line F = hij $\times$ m = 846 $\times$ m = *4*, so that m = 1 and line F = 846. But, line G = line H = hij $\times$ n = 846 $\times$ 9 = 7614, so that from lines E, F, and G, line E = line F + line G = 846 + 761 = 1607; that is, Ef = 7, Ee = 0, Ed = 6, and Ec = 1. Then, from lines C, D, and E,

$$\begin{array}{ll} \text{C.} & * \ * \ 4 \ * \\ \text{D.} & \underline{3\ 3\ 8\ 4} = 846 \times 4 = \text{hij} \times 4 \\ \text{E.} & \quad 1\ 6\ 0 \end{array}$$

Since the ten's column is 6 or 5, then h = 8; therefore, j = 6, n = 9, k = 1, and h = 8. Then hij = 846. Line D = hij $\times$ 4 = 846 $\times$ 4 = 3384; thus, Db = 3, Dc = 3, and Dd = 8. Line B = hij $\times$ k = 846 $\times$ 1 = 846; thus, Bb = 8, Bc = 4, and Bd = 6. Line H = hij $\times$ n = 846 $\times$ 9 = 7614; thus Gd = Hd = 7, Ge = He = 6, and Gf = Hf = 1. Then, Ad = Bd + Cd = 6 + 4 = 10; that is Ad = 0. Since line F = 846 and m = 1, then Fd = 8 and Ff = 6.

From above, line E = 1607, then Ed = 6, Ec = 0, and Ef = Af = 7; from lines C, D, and E, line C = line D + line E = 3384 + 160 = 3544. Thus, Cb = 3, Cc = 5, and Ce = Ae = 4. Finally, from lines A, B, and C, line A = line B + line C = 846 + 354 = 1200; thus, Ab = 2 and Ac = 0. The entire division is therefore,

```
           1 4 1 9
      ______________
 8 4 6)1 2 0 0 4 7 4
        8 4 6
       -------
        3 5 4 4
        3 3 8 4
        -------
          1 6 0 7
            8 4 6
          -------
            7 6 1 4
            7 6 1 4
            -------
```

39. the suspicious sailors no. #1

Let N number of bananas each sailor got at last division. Then $4N + 1$ is the number of bananas that the fourth sailor left. Since this is $\frac{3}{4}$ of the bananas left by the third sailor minus 1, then $\frac{4}{3}(4N + 1) + 1 = \frac{16N + 7}{3}$ is the number of bananas that the third sailor left. But, this is $\frac{3}{4}$ of the bananas left by the second sailor minus 1; thus, $\frac{4}{3}(\frac{16N + 7}{3}) + 1 = \frac{64N + 37}{9}$ is the number of bananas that the second sailor left. Then, this is $\frac{3}{4}$ of the bananas left by the first sailor minus 1; hence, $\frac{4}{3}(\frac{64N + 37}{9}) + 1 = \frac{256N + 175}{27}$ is the number of bananas that the first sailor left. Finally, this is $\frac{3}{4}$ of the original pile of bananas minus 1. Thus, if M is the number of bananas in the original pile, then, $M = \frac{4}{3}(\frac{256N + 175}{27}) + 1 = \frac{1024N + 781}{81} = 12N + 9 + \frac{52N + 52}{81}$.

Now, let $R = \frac{52N + 52}{81}$, then $81R = 52N + 52$, or $N = \frac{81R - 52}{52}$; thus, $N = R - 1 + \frac{29R}{52}$. Since N is an integer,

R must be a multiple of 52, so we may write $R = 52x$; thus, $N = \frac{81R - 52}{52} = \frac{(81 \times 52x) - 52}{52} = 81x - 1$ and $M = \frac{1024 \times (81x - 1) + 781}{81} = 1024x - 3$. If we choose $x = 1$, then $M = 1021$ bananas.

40. the suspicious sailors no. #2

Let $N =$ number of bananas each sailor got at last division. Then $5N + 1$ is the number of bananas that the fifth sailor left. Since this is $\frac{4}{5}$ of the bananas left by the fourth sailor minus 1, then $\frac{5}{4}(5N + 1) + 1 = \frac{25N + 9}{4}$ is the number of bananas that the fourth sailor left, which is $\frac{4}{5}$ of the bananas that the third sailor left minus 1; hence, $\frac{5}{4}(\frac{25N + 9}{4}) + 1 = \frac{125N + 61}{16}$ is the number of bananas that the third sailor left. Again, this is $\frac{4}{5}$ of the pile that the second sailor left minus 1, so that $\frac{5}{4}(\frac{125N + 61}{16}) + 1 = \frac{625N + 369}{64}$ is the number of bananas that the second sailor left; hence, this is $\frac{4}{5}$ of the pile that the first sailor left minus 1; therefore, $\frac{5}{4}(\frac{625N + 369}{64}) + 1 = \frac{3125N + 2101}{256}$ is the number of bananas that the first sailor left, which is $\frac{4}{5}$ of the original pile minus 1. Hence, $\frac{5}{4}(\frac{3125N + 2101}{256}) + 1 = \frac{15625N + 11529}{1024}$ is the number of bananas in the original pile. Thus, if M is the number of bananas in the original pile, then $M = \frac{15625N + 11529}{1024} = 15N + 11 + \frac{265N + 265}{1024}$.

Now, let $R = \frac{265N + 265}{1024}$, then $1024R = 265N + 265$, or $N = \frac{1024R - 265}{265}$; thus, $N = 3R - 1 + \frac{229R}{265}$. Since N is an integer, then $\frac{229R}{265}$ must also be an integer; hence, R must be a multiple of 265. Therefore, we may write $R = 265x$. Thus, $N = \frac{1024R - 265}{265} = \frac{1024 \times (265x) - 265}{265}$, or $N = 1024x - 1$, and $M = \frac{15625 \times (1024x - 1) + 11529}{1024} = 15625x - 4$. If we let $x = 1$, then $M = 15621$.

41. the scientific expedition

Suppose the truck starts out from point A and travels 216 miles to point B using $216 \div 12 = 18$ gallons of gasoline. At point B, the expedition deposits 54 gallons and returns to point A using up the remaining 18 gallons. This trip is repeated, so that another 54 gallons is deposited at point B. Point B, which is 216 miles from point A, now has deposits of $54 + 54 = 108$ gallons of gasoline. The truck then starts out again from point A fully loaded, travels to point B, and picks up the 18 gallons there which was used up to make the trip from point A to point B. Thus, there are 90 gallons now stored at point B. Then, fully loaded with gasoline, the truck proceeds to point C which is 576 miles from point A, but only 360 miles from point B. The truck then used up $360 \div 12 = 30$ gallons to get to point C. The expedition deposits 30 gallons at point C and returns to point B where the empty tanks of the truck are filled with 90 gallons from the gasoline stored there. Then, the truck proceeds again to point C, using 30 gallons, picks up 30 gallons there, and fills the tank again. At a full capacity of 90 gallons, the truck starts out from point C across the desert and travels the remaining $90 \times 12 = 1080$ miles to its destination. Thus, the expedition has used three full tanks of gasoline, or $90 \times 3 = 270$ gallons.